WEST COAST

Hazards Manual

a guide to local forecasts and conditions
3rd Edition

This Marine Weather Hazards Manual is dedicated to the memory of...

C.A. Casey
R.A. Cowlin
P. Malashewsky
J.C. Secord
S.T. Szczuka

—who lost their lives October 12, 1984, in the severe storm off Vancouver Island's west coast.

Environnement Canada Environment Canada

Canada

TABLE OF CONTENTS

CHAPTER ONE INTRODUCTION

CHAPTER TWO STORMS

CHAPTER THREE WIND

CHAPTER FOUR SEA STATE

CHAPTER FIVE WEATHER

CHAPTER SIX LOCAL HAZARDS

APPENDICES

TABLE OF CONTENTS

CHAPTER ONE

Introduction

Acknowledgements

The first edition of this manual was printed in 1987. The manual was well received and its basic structure has been retained. It was drafted by Seaconsult, Vancouver under contract to the Atmospheric Environment Service (A.E.S.), with editing completed by Marianne Pengelly and Eldon Oja of A.E.S. This second edition has been revised by Owen Lange, also of A.E.S. The graphic design and production was done by Jager Design Inc. Cover illustration, charts and maps by Harry Bardal.

We are grateful to many individuals representing the commercial fishing industry, commercial towing operations, Canada Coast Guard, the Department of Fisheries and Oceans, vessel Captains and recreational boaters whose input on what constitutes a marine weather hazard has greatly helped shape the revised edition of this hazards manual. We thank them for their time. We would also like to acknowledge AES Atlantic Region for their advice and co-operation.

Several minor changes have been made in this "3rd edition". The main content of the book is unchanged. The information about frequencies and observation sites of each forecast region in chapter six subject to frequent changes has been removed. For the information on weather observation sites obtain the latest copy of Environment Canada's "Mariners Guide". For information of radio frequencies see the Canadian Coast Guard's publication "Radio Aids to Marine Navigation" or on their web page www.pacific.ccg-gcc.gc.ca

Purpose of this Manual

Weather plays an important role for the mariner along the British Columbia Coast. Weather changes are often sudden and severe and represent a hazard both to the "old salt" with considerable experience at sea and to the sailing novice.

The Marine Weather Hazards manual is designed to help all mariners get the most out of Environment Canada's marine weather forecasts. It is not intended to be a textbook on marine meteorology but rather a clear, practical introduction to the local marine weather hazards on the West Coast.

This manual describes the different ways that storms approach the coast and how the strong winds, high seas and weather which result from these storms, are modified by various coastal effects. Chapter Six of the manual considers the specific areas which are known for marine weather related hazards.

Many specific marine weather hazards are mentioned but it would be impossible to list them all, even if they were known to the authors of this book. It is hoped that the mariner will be able to recognize the processes which cause the specific hazards and hence be able to anticipate other areas which might have the same conditions. In this way the waters should become a safer place for all those people who "go down to the sea in ships".

Marine Weather Publications

A new publication, "The Wind Came All Ways" follows the ideas of the Marine Weather Hazards Manual but takes a new approach and gives considerably more wind, wave and weather information for the waters from Juan de Fuca to Desolation Sound.

Marine Weather Video Services

A series of informative videos on marine weather along the British Columbia Coast are available to supplement the *West Coast Marine Weather Hazards Manual*. The videos are ideal for individuals and groups who want to learn more about local marine weather hazards. For more information on the video series, contact:

Environment Canada
Commercial Services
120 - 1200 West 73rd Avenue
Vancouver, B.C.
V6P 6H9

Visit our website for current marine forecasts at **www.weatheroffice.com**

Weather Services

Forecasts and warnings for all British Columbia's coastal waters are issued by marine meteorologists at the Pacific Weather Centre in Vancouver. Recent technology such as satellite imagery, automated coastal weather stations, weather buoys and computer data gathering and forecasting systems are all valuable tools used by the marine forecasters.

A full description of the marine weather services is available in the pamphlet "Mariner's Guide—West Coast Marine Weather Services". This pamphlet gives the marine forecast issue times, discusses the various parts of the marine forecast, and lists where weather information is available.

The Pacific Weather Centre, which is part of Environment Canada, issues marine forecasts for the following areas:

South Coast

Strait of Georgia
Howe Sound
Juan de Fuca Strait
Johnstone Strait
Queen Charlotte Strait
West Coast Vancouver Island North
West Coast Vancouver Island South

North Coast

Queen Charlotte Sound
Central Coast
Hecate Strait
Douglas Channel
Dixon Entrance East
Dixon Entrance West
West Coast Charlottes

Offshore

Bowie
Explorer

The map on the facing page shows the boundaries of the forecast regions as well as the location of all the coastal reporting stations (staffed and automatic), the marine weather buoys and the local weather offices.

Weather Services

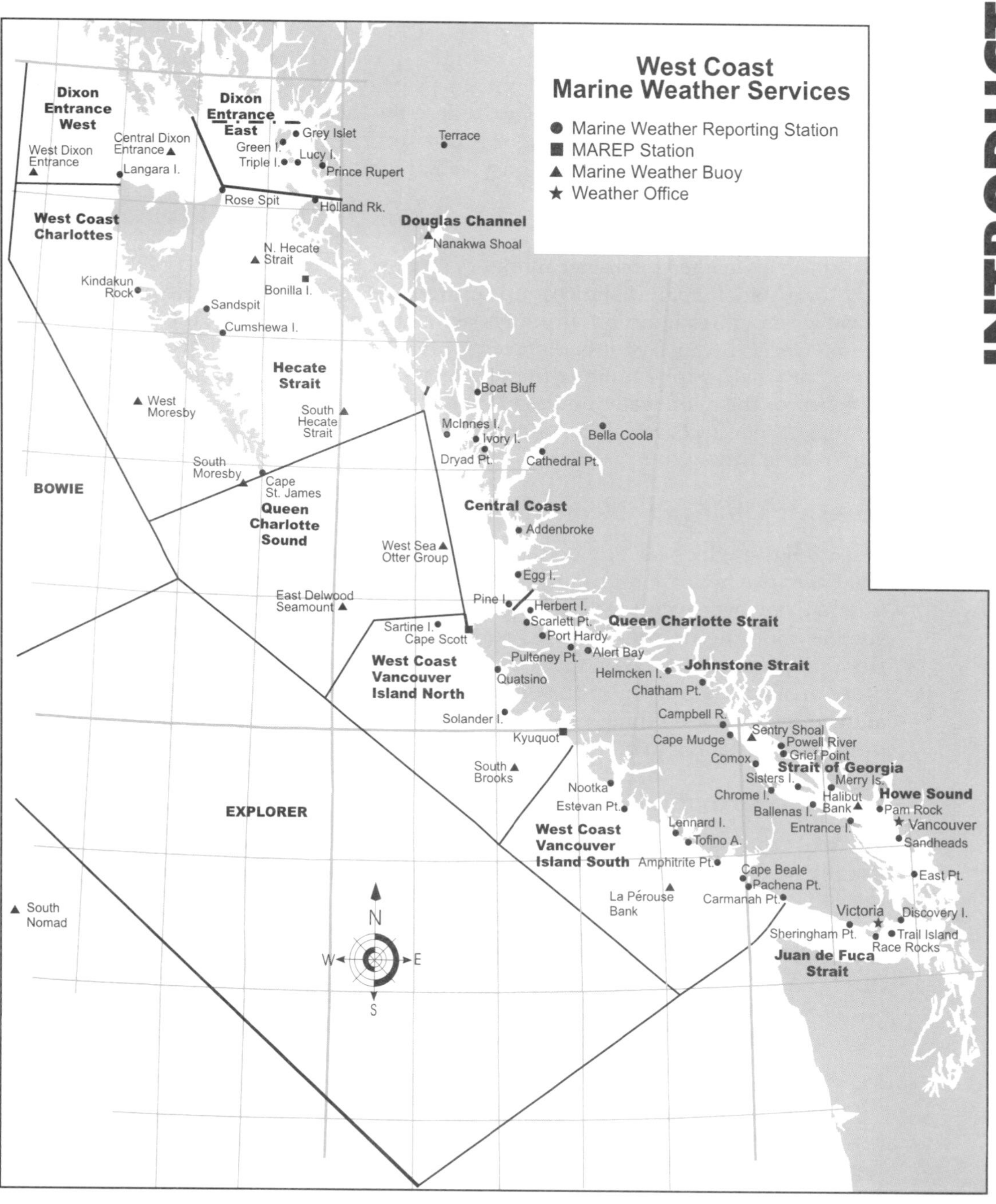

Marine Weather Forecasting

Weather along the West Coast can range from wildly dangerous over the open waters in winter to placidly calm in realms sheltered from summer breezes. It is the task of the marine weather forecaster to describe the expected weather conditions in all their variety. Three things make this task difficult: the different scales of weather systems, timing problems and forecast format limitations.

Different Scales

The marine weather forecaster monitors large scale weather systems such as highs and lows which span distances of hundreds or thousands of miles. The mariner, however, operates in an environment which may have significant weather differences over a distance of less than one mile. To bridge the gap between these two scales of weather phenomena the forecaster attempts to describe the winds and weather for the main water bodies of each marine forecast area. Consideration of the local weather effects which occur around every peninsula and behind each island is left to the mariner.

Forecast Timing Problems

The strongest winds usually occur just ahead of an approaching frontal system. The front can take hours and in some extreme cases, days, to cross a forecast region. The forecast, however, calls for the strong winds to arrive during a certain period of time, such as morning or afternoon. As some of the forecast regions are quite large the front may actually reach one part of the region right on schedule but may not clear other areas of the region until somewhat later. The mariner should learn to make allowances for these timing problems by watching the subtle changes in the sky conditions and listening to the "local weathers" from upstream locations.

Forecast Format Limitations

The most complete forecast is not always the most useful. Because the forecasts are usually made available to the mariners verbally, via the various marine radio broadcasts they cannot be overly long or detailed. Forecasters are often forced to leave out some of the conditions they may be aware of, simply for the sake of brevity. A forecast which tries to include every small wind change would be too lengthy.

Weather Check List

In this manual the reader will find descriptions of many aspects of marine weather as it would usually evolve, but the wise sailor knows that there is no substitute for a keen weather eye. The following check list summarizes the important features to consider.

1. **What is the present weather?**
   Listen to all the "local weather reports" in your area. Keep a "weather-eye" open.
2. **What forecast areas are important to you?**
   Make sure you listen to the right forecast.
   Listen to the forecast for the adjacent area for indications of upcoming weather.
3. **What is the weather summary?**
   Consider the location of fronts, highs and lows as described in the marine synopsis.
4. **What marine warnings are in effect or forecast?**
   Interpret these warnings according to the capabilities of your own boat and your ship handling abilities. As a simple guide the warnings can be considered as follows:

   small craft warning
   - be cautious

   gale warning
   - expect rough weather
   - very hazardous for small vessels

   storm and hurricane force wind warning
   - stay ashore or try to avoid the storm

5. **What is the forecast trend—worse, the same, or better?**
   Consider how long you will be at sea.
6. **What are the local marine weather hazards in your area?**
   Chapter 6 will help you identify these hazards.
   Learn how to make adjustments to the forecast for your own area.

Even if the weather forecasts were absolutely accurate and you have considered all the local hazards, you must still decide if the risks are acceptable to you.

Weather Maps

The weather map is as important to the forecaster as a compass is to the mariner. Through understanding the basic elements of a weather map the mariner can gain a different perspective on the movement and strength of the weather systems.

In order to read the weather maps in this manual, as well as the actual maps transmitted on marine facsimile machines, the basic features of a weather map are shown below and summarized opposite.

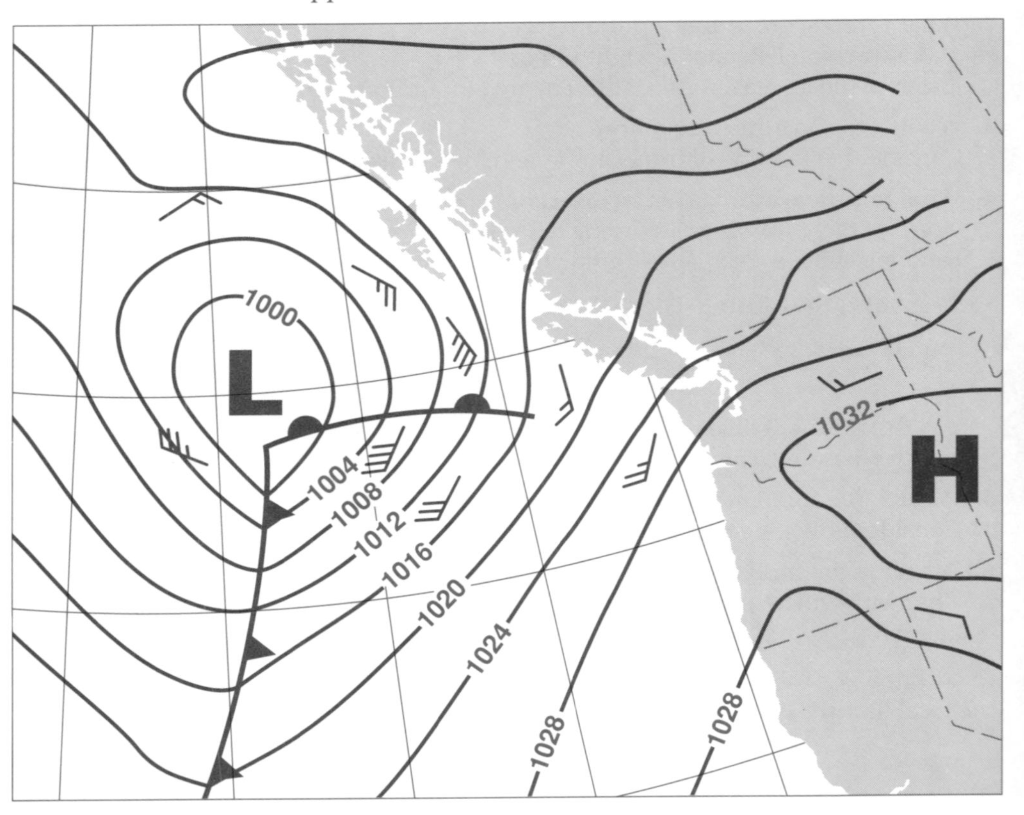

An example of a weather map, illustrating the different map elements.

Weather Maps

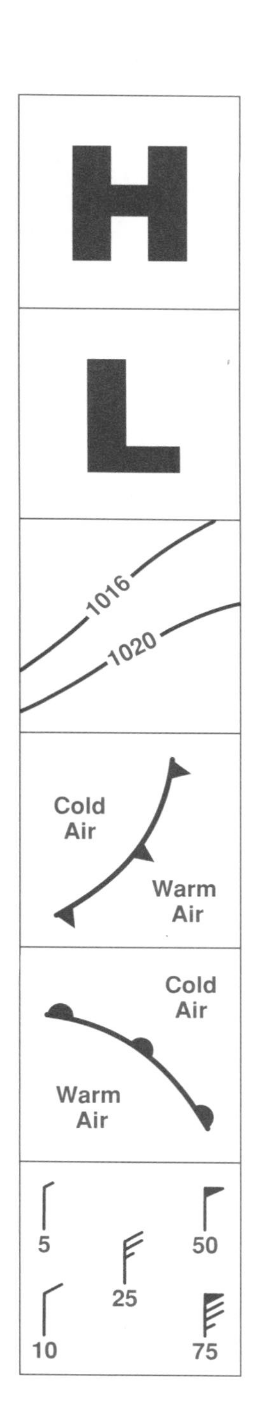

Centre of high pressure
The location of highest pressure. The pressure decreases in all directions out from the centre.

Centre of low pressure
The location of lowest pressure. The pressure increases in all directions out from the centre.

Isobars
Lines joining places of equal pressure. They are usually drawn at intervals of four millibars. The closer the isobars, the stronger the wind.

Cold front
The leading edge of an advancing cold air mass which usually moves southeastward. The "icicles" on the front point from the cold to the warm air.

Warm front
The trailing edge of a retreating cold air mass which usually moves eastward. The "raindrops" on the front face into the cold air.

Wind speed flags
The arrow flies with the wind, indicating the direction from which the wind is blowing. The wind speed, in knots, is given by the number of barbs and/or flags on the shaft.

MAREP

The *Mar*iner *Rep*orting program (MAREP) is how you—the mariner—can be involved in the process of weather forecasting.

Timely weather reports can be life saving in severe weather situations. Environment Canada's Atmospheric Environment Service (AES) participates in several cooperative programs with the marine community to increase the number of weather observations from mariners at sea. These reports assist the forecasters at the Pacific Weather Centre in preparing forecasts and warnings. In addition, other skippers can receive this vital, first-hand information to supplement the regular marine weather radio broadcasts.

In one program, reports are transmitted via VHF channel 69 to shore-based AES MAREP operators at Bonilla Island, Cape Scott and Kyuquot. These operators are well known to the local marine communities and have a direct communication link with the Pacific Weather Centre. Marine forecasts and warnings are then relayed by the MAREP operator to ships within VHF radio range.

Cape Scott and Bonilla Island Marep stations presently operate from October 1 to April 30. The Kyuquot Marep station operates from May 1 to September 30. Marep reports can be sent at any time of the year in critical weather situations. To check on changes to the Marep or other marine programs visit the marine page of www.weatheroffice.com

Another cooperative program between the Atmospheric Environment Service, the Canadian Coast Guard Radio stations and the marine community has participating vessels call in MAREP reports via VHF channels 26 and 84 to the nearest Coast Guard radio station who in turn relay the data to the Pacific Weather Centre. These reports may be broadcast over the Coast Guard marine radio.

All observations are useful, but the most important reports are those which tell of weather conditions that have suddenly worsened or those that greatly differ from the forecast. The observations can be as brief as a simple estimate of the wind. A complete observation from the at-sea reporter might contain the following details:

Location	Five miles west of Tofino... near Learmonth Bank... rounding Cape Cook, 3 miles SW of Solander... An approximate position is enough.
Wind	An estimate of speed and direction
Sea	General sea state conditions.
Visibility	The distance it is possible to see.
Weather	Any important weather features such as: thunderstorms, waterspouts, sharp wind directional changes...

MAREP

For those with limited experience, making an observation at sea of the wind speed and the height of the sea can be quite a challenge. (Pictures of four sea state conditions are found on pages 60–61.) The following table is provided as a beginner's guide to the relationships between the strength of wind and the associated seas. The descriptions below are a modified version of that originally given by Admiral Beaufort in the early nineteenth century.

Wind Description	Wind speed (knots)	Effects of Wind at sea	Additional Information	Probable Wave Height (metres)
Light	0–11	Wavelets form. A few crests break near 10kt.	Sails fill above 5kt. Light flags extended.	0–1/2
Moderate	12–19	Waves lengthen. Frequent white horses.	Wind pressure felt on face. Good sailing breeze.	1/2–2
Strong	20–27	Large waves form. Extensive foam crests. Some sea spray.	Heavy flags flap vigorously. Small sailboats in difficulty.	2–3
Near gale	28–33	White foam from breaking waves blown in streaks.	Large sailboats reduce sail. Small boats remain in harbour.	3–5
Gale	34–47	Foam streaks become very dense. Wave crests begin to topple.	Heavy flags fly straight out and whip from the hoist.	5–8
Storm	48–63	Very high waves. Sea has white appearance. Visibility affected.	Large sailing ships close reefed or running with lower topsails.	8–12
Hurricane Force	64 or more	Air filled with foam and spray. Sea entirely white. Visibility seriously impaired.		above 12

Note: The probable wave height is a guide to what may be expected in the open sea, remote from land. In enclosed or sheltered waters, the wave heights will be smaller.

WINTER STORMS
- a) Gulf of Alaska lows
- b) Coastal lows

ARCTIC OUTBREAKS

SUMMER STORMS
- a) Summer fronts
- b) Lee trough
- c) Stratus surge

THUNDERSTORMS

CHAPTER TWO

Storms

Winter Storms

Marine weather hazards are caused primarily by weather systems which affect the coast and create strong winds, high seas and heavy rains. As the term "storm" is often used by mariners to describe these severe weather events it has been used as a general title for this chapter.

From approximately early October to April, **winter storms** occur frequently over the northeast Pacific Ocean. On average, 10 to 15 storms will affect the west coast each month. The winds generally rise to gale force and often to storm force. Hurricane force winds usually occur several times during the winter at some of the more exposed locations.

The transition from the quiet days of summer to the severe weather of winter is often not a gradual change but can be very dramatic. Storms near the autumn equinox often mark this transition. On September 28, 1990 there was a classic equinoctual storm. At the West Dixon Entrance buoy northwest of the Charlottes the winds rose from light to 54 knots and the seas built from 2.5 metres to 11.4 metres in just 10 hours.

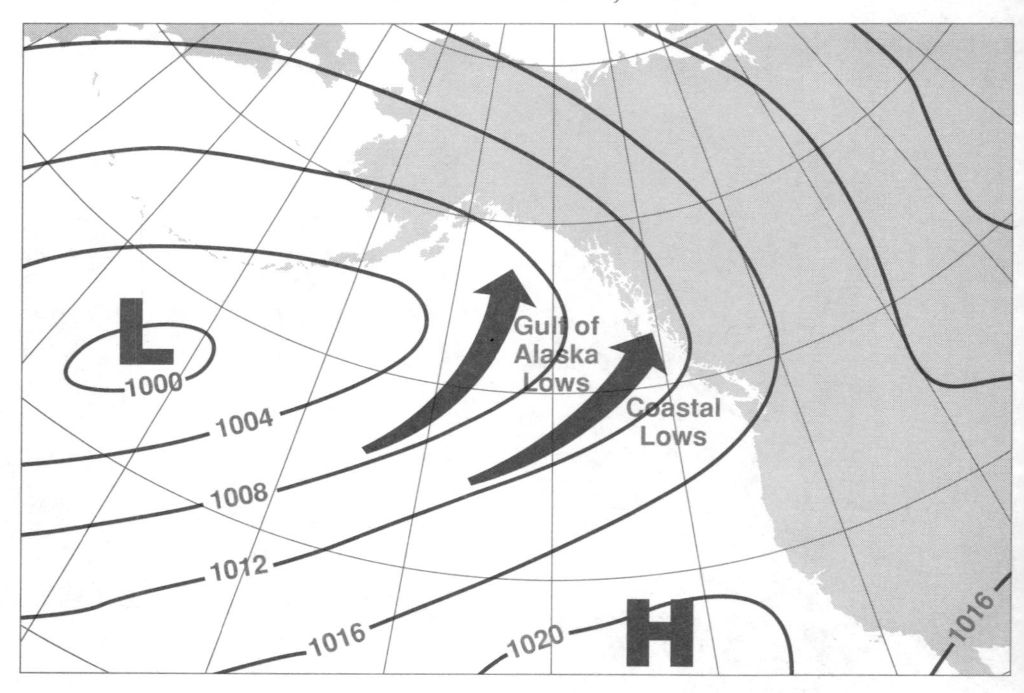

Principal winter storm tracks are superimposed on the January mean sea-level pressure pattern. (Pressure values in millibars)

Winter Storms

Most of the winter storms are **Gulf of Alaska lows** that remain well offshore and track northward into the Gulf of Alaska. A frontal system will normally move away from the low and pass southeastward over the coast. The strongest winds and the heaviest seas generally occur just ahead of the front. The front usually weakens as it crosses the southern coastal waters.

A second type of winter storm, called a **coastal low,** develops rapidly quite close to the coast and moves directly across the B.C. waters. A coastal low can be very hazardous because of its speed of movement and development. Another dangerous aspect of the coastal low is that the winds vary considerably around the low. The strongest winds and the highest seas usually occur to the south of the low.

On a few occasions during the winter a low will combine the paths of both the Gulf of Alaska low and the coastal low. In this situation the low first moves into the Gulf of Alaska and begins to weaken. The front sweeps over the coast giving gale to storm force winds. After a time the low sitting over the Gulf of Alaska, begins to drift southeastward toward the B.C. coast and strengthens once again. The strong winds which swirl around the low are then carried directly over the coastal waters. These lows are sometimes referred to as **polar lows**.

The map on the facing page shows the average tracks, or paths which the different types of lows usually follow. Actual storm tracks can vary from those shown.

At Cape St. James, the maximum recorded wind speed between 1957 and 1983 was 95 knots. This observation was made in October, 1963.

a) Gulf of Alaska Lows

A Gulf of Alaska low usually forms south of the Aleutian chain as a frontal wave between cold northern air and warmer air to the south. Such a wave may travel a considerable distance eastward before it begins to take shape as a low pressure system. Once the low starts to develop, pressures fall rapidly. At the same time the low pressure system increases in size and wind speeds rise. The air flow is always counterclockwise around the low.

Winter Storms

The low often deepens to 970 millibars, or lower, while moving northeast at 35 to 40 knots. The low usually reaches its lowest central pressure over the Gulf of Alaska. The front that extends southward from the low centre, sweeps over the coastal waters bringing with it southeasterly gale-to-storm-force winds, rain and heavy seas. With the most active fronts the winds can briefly rise to hurricane-force strength (64 knots and higher) at places such as Solander Island, on the northern end of Vancouver Island. As the low moves further north over the Alaskan coast and the front crosses the B.C. coast the storm enters its mature stage and begins to weaken.

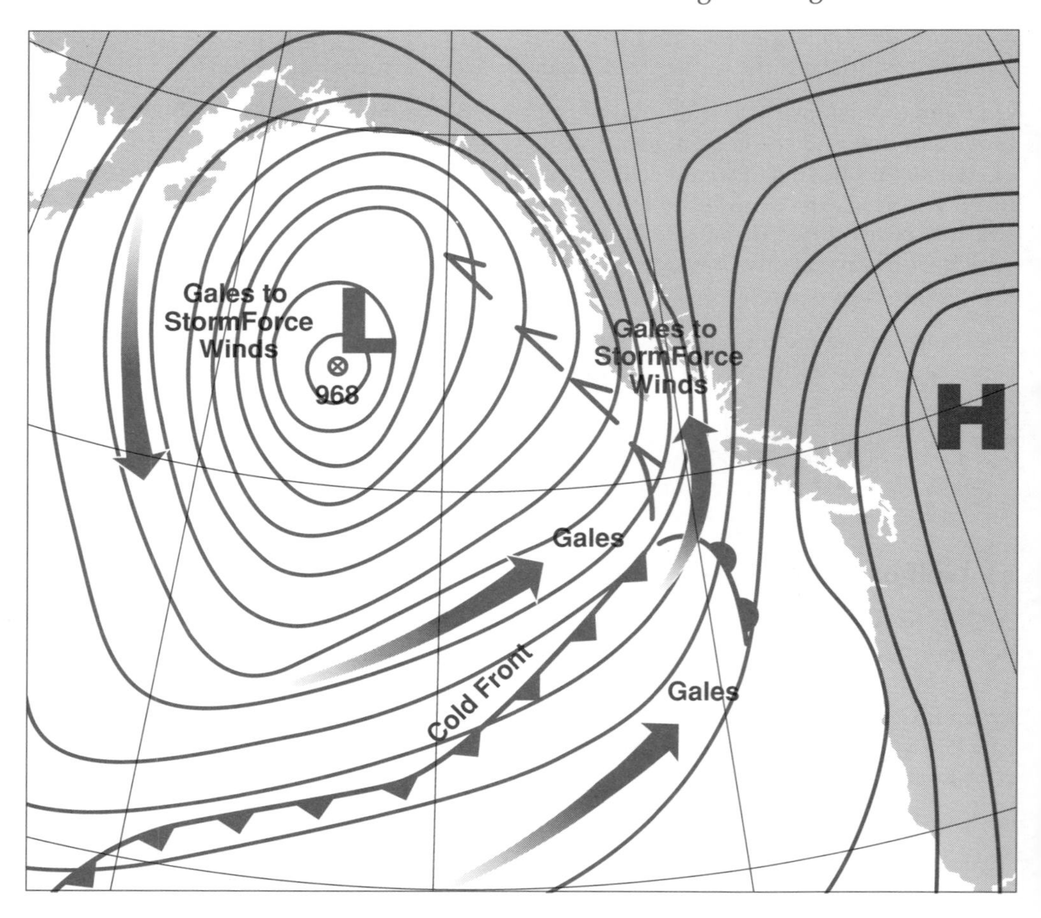

This sea-level pressure pattern indicates a Gulf of Alaska low with the associated frontal system approaching the B.C. coast. Gale-to-storm-force southeasterly winds are often found just ahead of the front with gale-to-storm-force northerlies just to the west of the low. (Pressure values in millibars)

Winter Storms

Ahead of the front, gale-to-storm-force southeasterly winds often build the seas to 9 metres over the northern waters and to the west of Vancouver Island. In the most intense storms the seas can reach 10 to 12 metres. Over the inner waters the seas are much lower due to sheltering from coastal land masses.

After the front passes over the coast the winds usually decrease, the visibility improves and the seas diminish. Heavy swells, typically from 4 to 6 metres, arrive at the coast up to 12 hours after the front has passed. These waves were formed with the strong west or northwest winds on the western flank of the low.

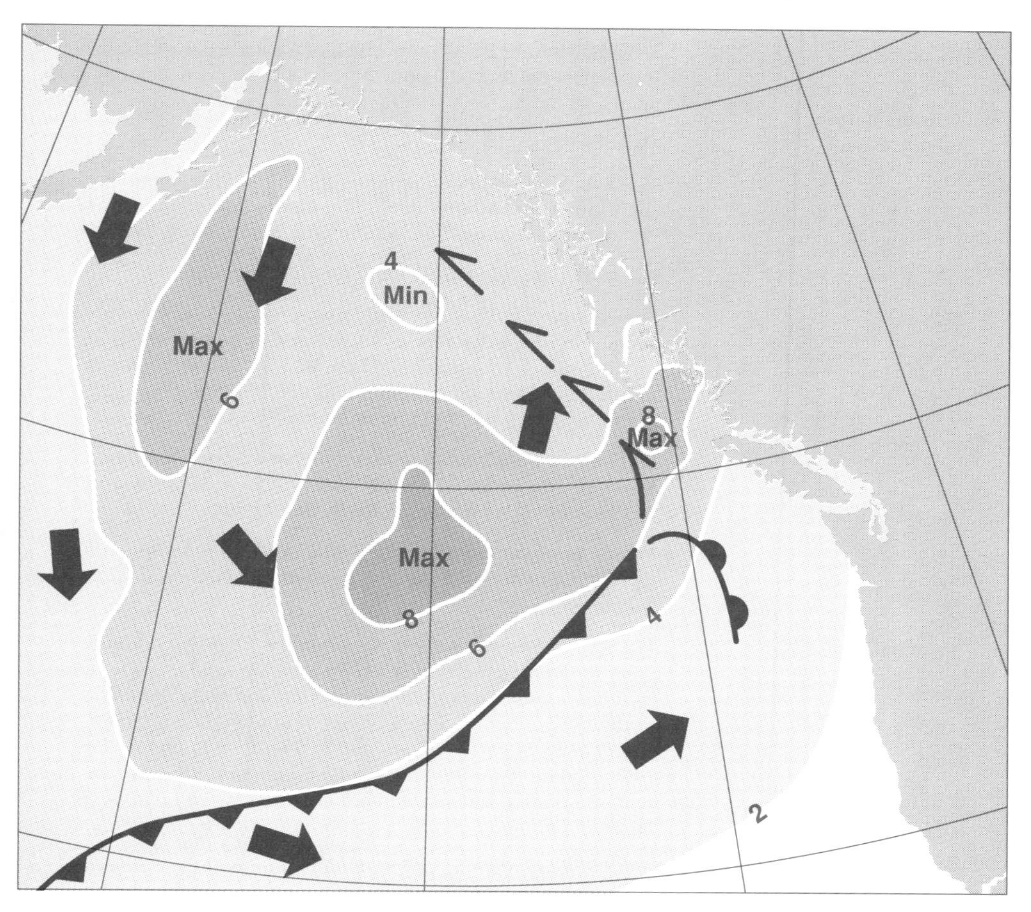

A typical pattern of wave heights (in metres) found over the eastern Pacific with a deep, Gulf of Alaska low. The arrows show the direction of movement of the seas.

Winter Storms

As the front approaches a vessel at sea the observed sequence of weather is quite regular and provides a number of warning signs for the observant mariner. Although the time of weather changes will vary from storm to storm, the following weather description is typical of an approaching front.

An actual weather sequence at Cape St James associated with a Gulf of Alaska low.

Condition: 968 millibar (mb) low in central Gulf of Alaska, front approaching the coast at 35 to 40 knots.

Hours Before Front Passes	Weather
21 h	Winds are light westerly, scattered clouds and barometer rising slowly. Gradually increasing cirrus and cirrostratus cloud provides an early indication of an approaching front.
17 h	Cirrus clouds widespread, winds shift to light SE.
15 h	Sky overcast with thickening cirrostratus clouds, winds remain light southeasterly but barometer begins to fall at about 1 mb/h (millibars per hour). Later, clouds lower to altostratus and winds increase to SE 10 to 15 knots providing a good warning of an impending storm.
11 h	Skies heavily overcast with altostratus and lower cloud, rain has begun and winds increase to SE 18 to 20 knots. Pressures continue to fall, a little more rapidly than before.
5 h	Winds increase to gale force S–SE. Visibilities lowered to three miles in rain and fog with pressure now falling at 1.5 to 2.0 mb/h.
Frontal Passage	Within 5 to 6 hours of the front passing, S–SE gales continue. Near the front, winds reach S–SE 50 gusting to 60 knots with visibility lowered to 1.5 miles in rain and fog and to 3/4 mile at the front. Pressure falls at 3 mb/h until the front passes. After the front passes winds shift to SW 20 to 25 knots with little change for the following 12 hours, accompanied by widely scattered showers and barometric pressure increasing at about 1 mb/h.

These weather signs often give 6 to 8 hours warning of gale force winds in Gulf of Alaska lows.

Winter Storms

b) Coastal Lows

Coastal lows usually intensify very quickly just before they move over the B.C. coastal waters. They can change from a very weak system into a severe storm in as little as 9 hours. The pressure will often fall 3 to 4 millibars every hour just ahead of the storm as it moves rapidly toward the coast.

Lows which do develop in such a rapid, or explosive manner, are referred to by the forecasters as "bombs".

The strongest winds are usually to the east or southeast of the low, just ahead of the associated front, and may reach southeasterly 70 knots. Gusts up to 100 knots occasionally occur in the most severe storms. Often, a second band of strong winds occur behind the front in an area south of the low pressure centre. Here, winds may range up to 60 or 65 knots from the west to northwest. The lightest winds often occur to the north and northeast of the low. Due to the varied strengths of winds around a coastal low, the exact track of the low can be extremely important for the mariner.

Lows that produce an extensive area of storm force winds have central pressures near 970 millibars or lower. Gales generally occur with lows having central pressures of 980 to 990 millibars.

The coastal lows often move through Queen Charlotte Sound or over the Queen Charlotte Islands. Occasionally a low will move eastward, passing just south of Vancouver Island. The lows which follow this southern track can bring very strong winds and heavy rains to the heavily travelled waters near Vancouver and Victoria.

Winter Storms

High winds with seas reaching 5 to 8 metres in exposed locations plus poor visibility in rain and fog precede the front of a coastal low. These conditions are very hazardous. Peak seas of 9 to 12 metres can be encountered behind the front, about 100-150 miles south of the low. The lowest seas, perhaps only building to 3 to 4 metres, would normally occur in the northeast quadrant ahead of the low.

As there is such sharp variation in the heights of the sea around the low it is important to be aware of the exact track of the low. A mariner who is to the south of the track of a coastal low would experience much worse conditions than one who is to the north of its track.

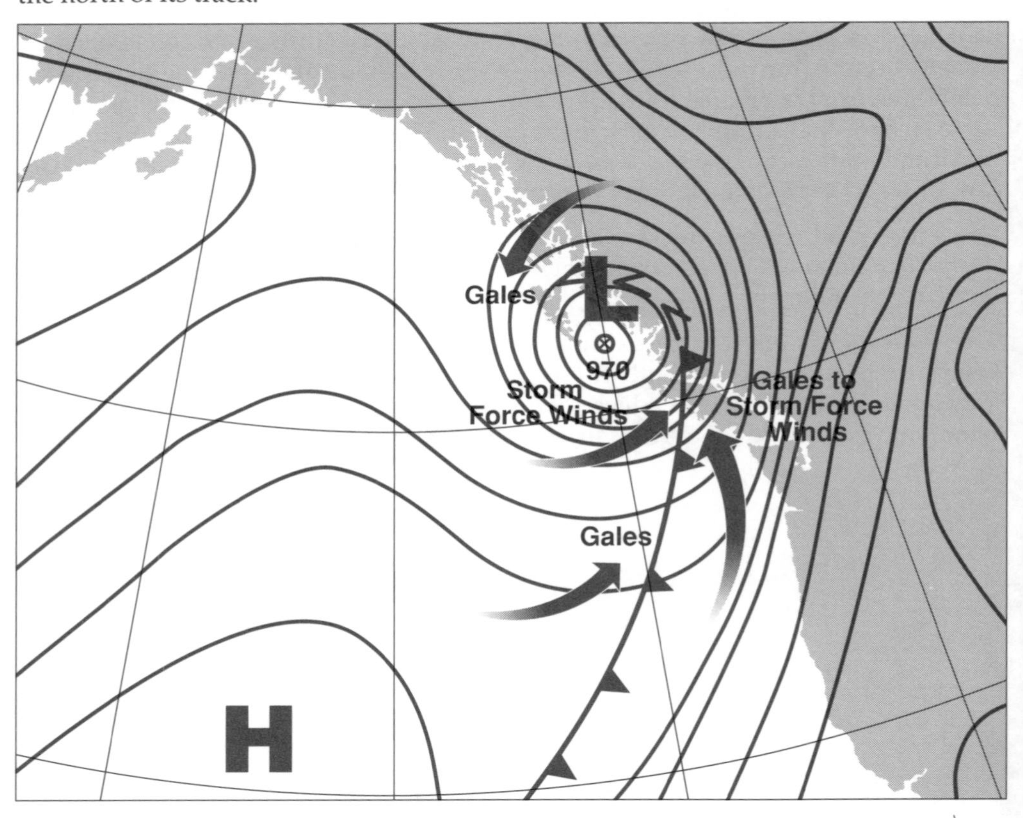

A typical sea-level pressure pattern for a coastal low and associated frontal system, with the winds superimposed.

Winter Storms

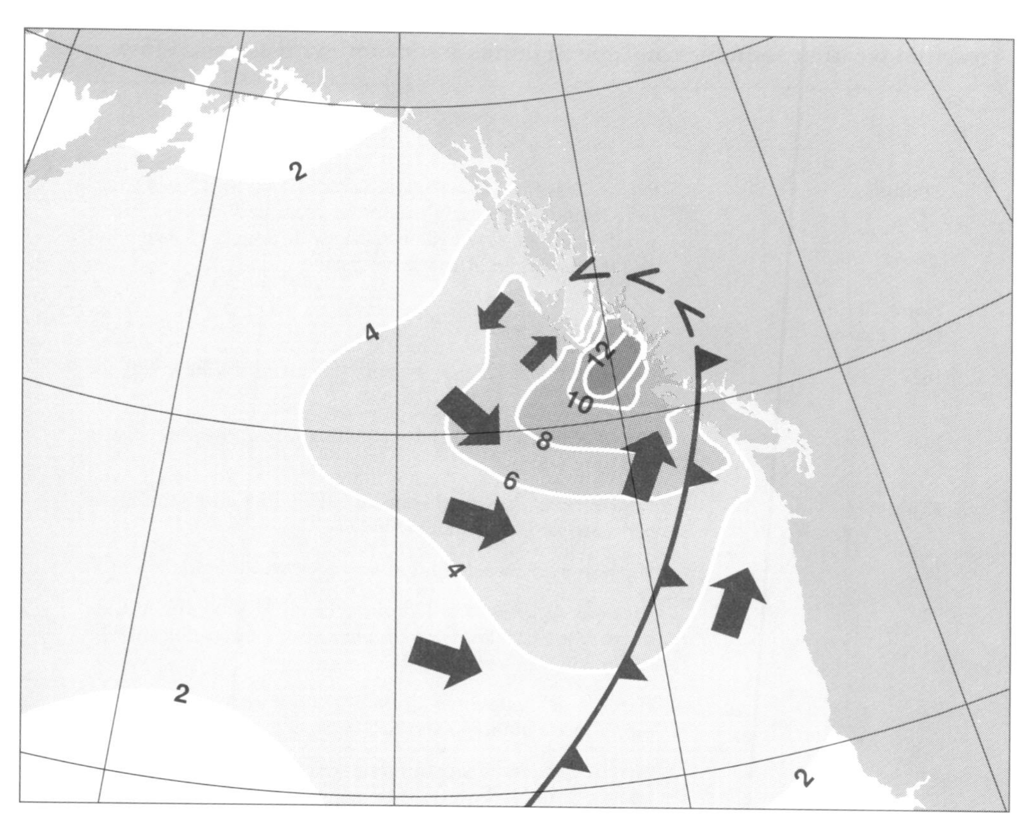

A coastal low with its associated front has just crossed the north coast. Wave-height contours are shown in metres and direction-of-wave motion is shown by arrows.

Winter Storms

An actual weather sequence at Cape St James associated with a coastal low.

Condition:	976 millibar (mb) low deepening rapidly off the Queen Charlotte Islands moving NE onto the coast at 35 knots. Strong winds and high seas in Queen Charlotte Sound, Hecate Strait and the north end of Vancouver Island.
Hours Before Front Passes	**Weather**
16 h	Cirrus clouds visible on the horizon and spreading across the sky.
14 h	Cirrus thickens, overcast with altostratus. Pressure steady and light winds.
13 h	Sky overcast. Pressure begins to fall about 1 mb/h (millibars per hour) and winds shift to light SE.
11 h	Sky heavily overcast. Rain commences, winds light SE.
9 h	Pressure begins to fall rapidly (3 mb/h) in rain. Winds increase to SE 20 knots providing a strong indication that the storm will be severe.
7 h	Winds reach gale force. Pressures continue to fall, now at 4 mb/h and visibility lowers to 3 miles in fog and rain.
4 h	Winds increase to storm force—SE 50 knots gusting to 65 knots. Storm conditions continue for another four hours until the cold front passes.
Frontal Passage	The cold front passes. Rain ends, winds S-SE 57 knots gusting to 75 knots. Following the front, winds increase to SW 65 knots gusting to 80 knots in light showers. Seas peak at 10 to 12 metres in Queen Charlotte Sound just after the front moves through. Pressure rises at one to 2 mb/h and winds continue above gale force for another 9 to 10 hours shifting to the NW. Seas decay steadily becoming swell of 4 to 6 metres in Queeen Charlotte Sound 15 to 18 hours after the front crosses the coastal waters. Waves in Hecate Strait are not as large due to sheltering, but may be hazardous if steepening on ebb tide currents.

Winter Storms

A satellite photograph shows the cloud pattern of a coastal low as the cold front reaches the mainland coast in Northern B.C. and crosses the north end of Vancouver Island. The low centre is located about 60 miles southwest of the Queen Charlotte Islands. Maximum waves may occur just south of the low centre.

Arctic Outbreaks

During winter, very cold air forms an area of high pressure over Alaska and Arctic Canada. This cold arctic air spreads over northern B.C. and occasionally flows into the southern interior of the province. When the layer of cold air thickens, it can cascade down the mountain passes producing very strong winds and some of the most treacherous weather through the coastal inlets.

When the arctic air becomes firmly entrenched over B.C., gale or storm force winds may persist through the inlets without respite, for several days. Along the South Coast, the arctic outflow occurs less frequently. But at least once a year, the arctic air manages to push westward, sometimes reaching the west coast of Vancouver Island.

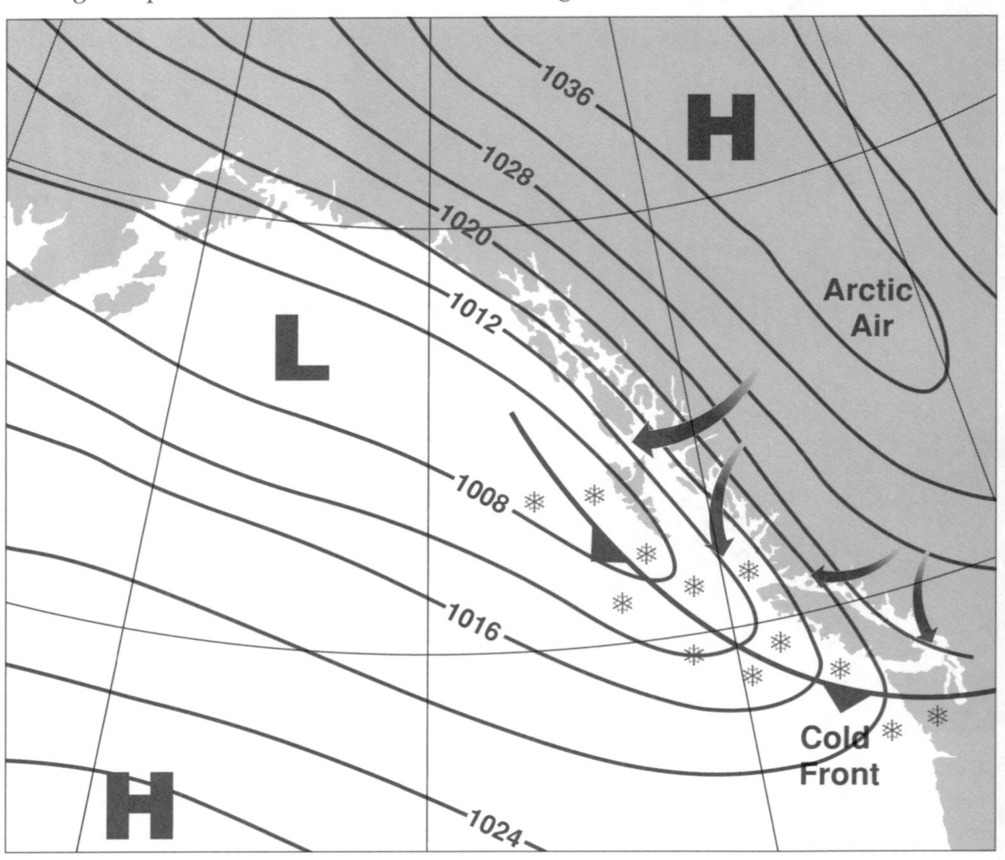

A ridge of high pressure builds over the province as the cold, arctic air flows into the Interior. This pushes the cold front out onto the coast. Outflow winds of 60 knots or more can occur through the mainland inlets and near the mouth of the inlets.
(Pressure values in millibars)

Arctic Outbreaks

There are a number of marine hazards associated with an **arctic outbreak** on the coast—gale or storm force winds, severe icing conditions and poor visibility. When they all occur together, as they often do, marine conditions become extremely dangerous. The arctic front is at the leading edge of the cold air. Its arrival is the first indication of the deteriorating conditions to follow. Weather on the coast preceding the front gives little sign of the impending changes. The barometer is often steady and the weather may seem quite settled. When the arctic front arrives it is usually accompanied by snow or snow showers. As the front passes, the temperature drops sharply, the barometric pressure rises rapidly and strong northerly outflow winds begin. The band of snow will move westward with the front and as drier air moves into the area, flurries end and skies clear. Clear skies will persist along the mainland coast throughout the period of outflow winds. However, the cold, dry air flowing over bodies of water such as the Strait of Georgia and Hecate Strait pick up enough moisture to give snow showers along the east side of Vancouver Island and the Queen Charlotte Islands.

In the arctic air behind the front, outflow conditions persist. Strong winds funnel down mainland inlets at speeds often up to 60 knots and occasionally rising as high as 100 knots. Side tributaries from the main inlets can also have strong winds and where a major side valley joins the main inlet, chaotic conditions are found.

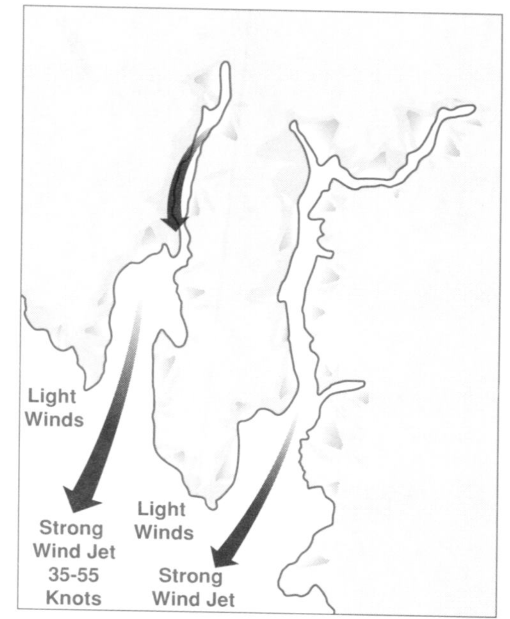

As the strong winds flow out of the mouth of the inlets they continue for some distance but gradually weaken when the flow is no longer confined by the narrow inlet. Extreme caution is advised when crossing coastal inlets during an outflow situation as the winds could increase abruptly in a narrow band near the mouth of the inlet. Because of the localized character of the winds, many reporting weather stations are not able to monitor these outflows.

Arctic Outbreaks

Perhaps the most widely known arctic outflow wind is the **Squamish.** It blows down Howe Sound and takes its name from the town at the head of the inlet. (In fact, the term Squamish is often applied to strong outflow winds in other coastal inlets.) There is often a sharp difference between the strong gales through Queen Charlotte Channel, just south of Bowen Island, and the generally light winds elsewhere. The Point Atkinson Lighthouse, located just to the east of Howe Sound on Burrard Inlet, often reports a light easterly wind during an arctic outflow. Just a short distance west, however, there will be northeast gale force winds.

Snow showers often accompany the initial surge of cold air. As the outflow persists, the air gradually becomes drier and the weather turns clear and cold. As the air travels out of Howe Sound and over the Strait of Georgia it gradually picks up moisture. Snow showers can be frequent and quite heavy along the east coast of Vancouver Island with significant amounts of snow in the Duncan-Nanaimo area.

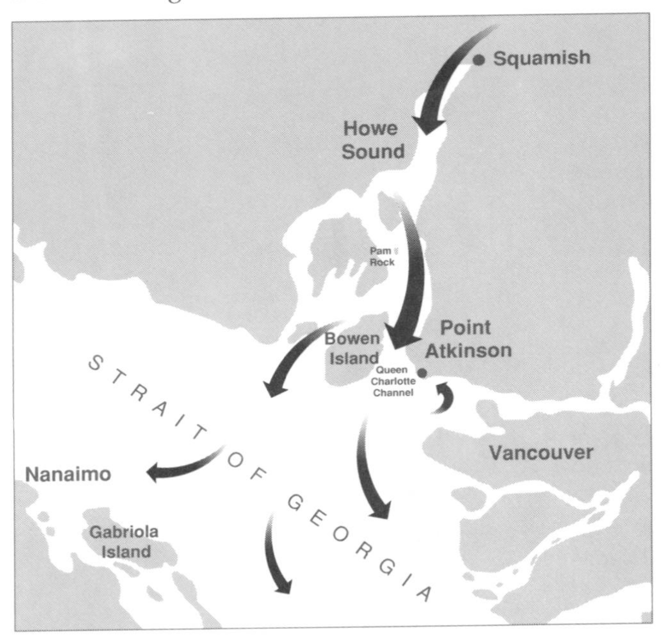

When cold air moves out over the mainland inlets, sea smoke is formed. Air temperatures must be well below freezing (minus 10° Celsius or less) for this to happen. In most cases sea smoke is confined to a metre or two above the water surface, but at times it can be much deeper and reduce visibility to near zero.

Another kind of fog has been noted under the most severe outflow conditions. It occurs when extremely strong gusty winds churn the water into a frenzy. The smallest spray particles freeze in the air and reduce the visibility. This type of ice fog has been reported with winds gusting to 80 knots. The larger particles of freezing spray are the main contributors to superstructure icing.

Arctic Outbreaks

The end of an arctic outbreak occurs when the cold air is forced away from the coast by the arrival of a Pacific storm. Unfortunately, this is accompanied by even more unfavourable weather.

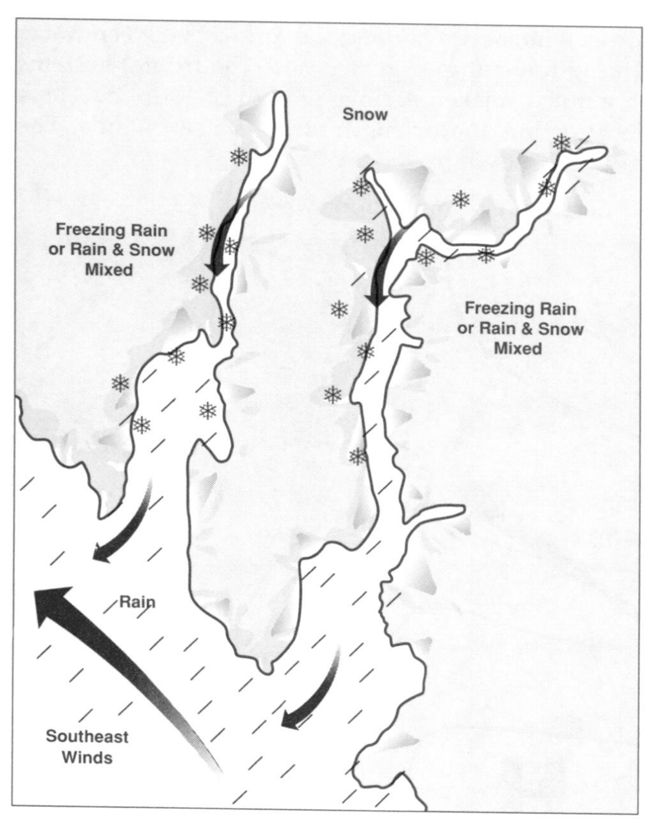

As the storm approaches it brings warmer air from the Pacific. The precipitation with the associated front would be either rain or mixed rain and snow. Some cold air may remain trapped in the inlets, and is not immediately warmed up. The northeasterly winds through the inlets gradually ease as the southeast winds ahead of the front strengthen. Often, rain will be reported at the mouth of the inlets while snow continues further up the inlet. Freezing rain can also occur in the inlets until the cold air is fully scoured out by the approaching warmer air. Freezing rain, or mixed rain and snow can significantly reduce radar efficiency.

Summer Storms

During the summer months, from May to September, the semi-permanent, Pacific high-pressure area off California strengthens and moves further north. As a result, the storm track is deflected by the high into the northern Gulf of Alaska.

Summer storms are not as intense as winter ones because the contrast between warm and cold air is not as strong during the summer. As a result, the frontal systems which approach the B.C. coast are much weaker. A ridge of high pressure develops near the coast and controls the direction and strength of the coastal winds. The winds will occasionally rise to gale-force strength.

The prevailing winds during this period are from the northwest.

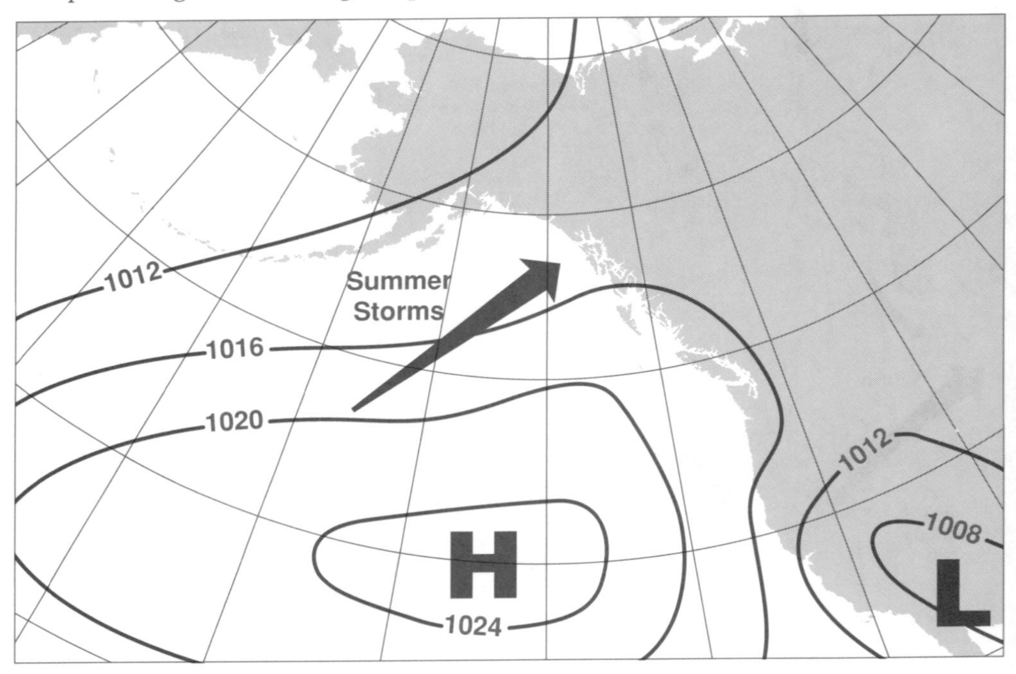

A normal Summer Storm track shown on the July mean sea-level pressure pattern. (Pressure values in millibars)

Summer Storms

a) Summer fronts

The winds in advance of the front are generally from the south or southwest. These winds will back into the southeast only in the waters very close to the coast. The front is usually accompanied by a narrow band of cloud and light rainfall. As the front continues southeast over Vancouver Island, the rain area often disappears and the clouds begin to dissipate.

Behind the front however, pressure rises are fairly strong—0.5 to 1 millibars per hour—as the ridge of high pressure attempts to rebuild along the coast. This produces strong northwest winds following the frontal passage. The strongest northwesterly winds are often reported in the Strait of Georgia where the airstream is funnelled between the mountains of the mainland and Vancouver Island.

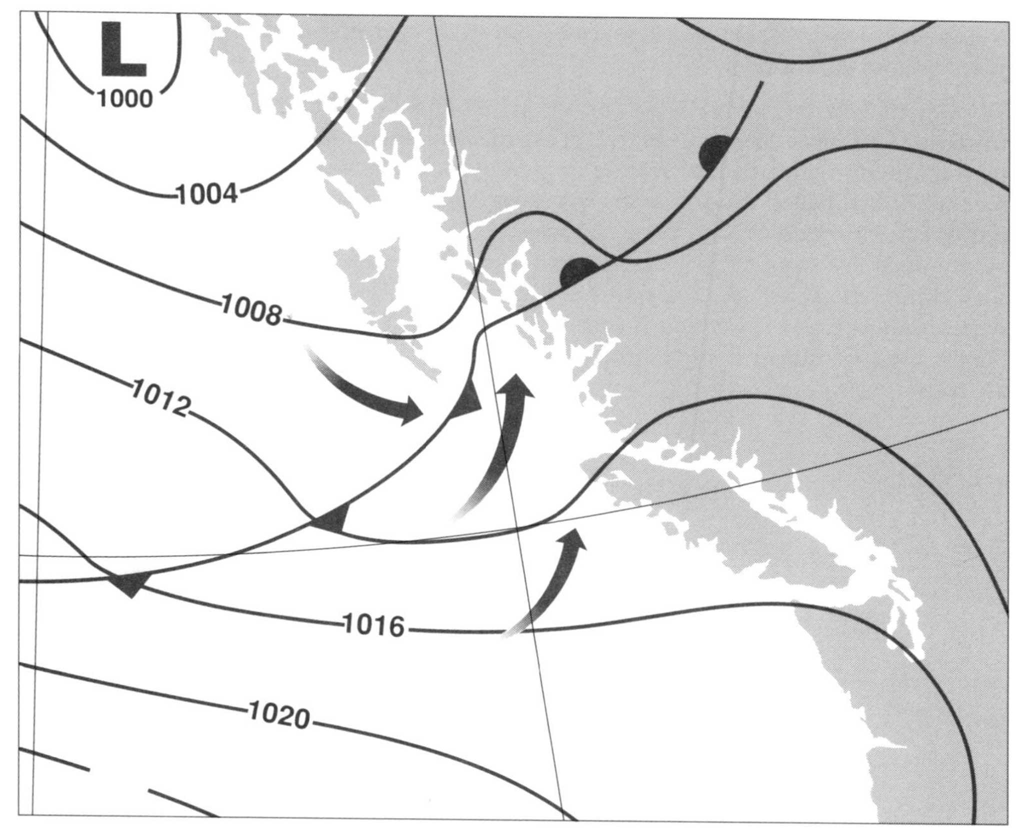

A typical summer pressure pattern shows a front crossing the coast with an indication of the winds near the front. (Pressure values in millibars)

Summer Storms

In spring or early summer, development of strong winds is possible since the fronts still retain some of the strength of winter storms. In fact, one of the strongest winds recorded at Sandheads was northwest 53 knots in March 1975.

Mariners can easily be caught unaware when this situation develops as there are few indicators of the impending changes. The front may have little, if any, cloud associated with it. Winds in advance of the front may decrease slightly or even remain light northwesterly, with steady barometric pressure. Behind the front, pressures begin to rise very rapidly with a sudden increase in wind to gusty, northwest gales.

Pressure rises of 1 millibar per hour or greater, at weather reporting stations located north of the front, may provide some warning of expected changes along the South Coast. However, winds at these stations would not likely reflect the strength of northwest winds that would develop later in the Strait of Georgia. Close monitoring of the weather forecasts is necessary for an early warning of these potentially dangerous wind conditions.

Once or twice a year, there is a situation which develops in quite a different manner from most summer fronts. In March or April, when the interior is enjoying unusually high temperatures, a frontal system may move onto the south coast followed by cold air that has travelled around a low pressure centre offshore. As the cold air reaches the coast, extremely strong pressure rises of 3 to 6 millibars per hour are recorded. Winds increase from 10 to 15 knots, to southerly 50 to 60 knots within an hour and then diminish as rapidly. The entire episode lasts only a few hours but the suddenness and severity with which it occurs means that it is potentially very dangerous. Again, close monitoring of weather forecast information is necessary if precautions are to be taken.

Summer Storms

Sudden gales in summer storms can build heavy seas with reduced visibility.

Summer Storms

b) Lee trough

Strong winds in summer frequently arise not from the intense pressure systems which move onto the coast but from more subtle local effects which take place all across B.C. The usual summer circulation pattern is a high pressure area over the eastern Pacific and a trough of low pressure in the southern interior. This trough is a thermal (heat) trough which forms after prolonged heating.

During the summer the thermal trough over the southern interior will cause an easterly flow of winds aloft which forms a **lee trough** just west of the coastal mountain ranges. This trough often lies over Georgia Strait and to the west of Vancouver Island. When the trough is particularly strong and the ridge of high pressure builds offshore, a band of strong, northwesterly winds, is formed on the western edge of the trough away from the coast. These strong winds, sometimes rising to gale force, are often recorded at Solander Island, just off the Brooks Peninsula. The winds adjacent to the coast however, as well as those through Juan de Fuca Strait, are usually light easterlies of 15 knots or less. Some of the warmest temperatures recorded in this region occur in this type of weather pattern. Even on the west coast of Vancouver Island temperatures can exceed 25° Celsius.

The lee trough may persist for several days until a disturbance from the Pacific moves eastward, forcing the ridge of high pressure closer to the coast. When this happens the lee trough breaks down and the northwest winds spread onto the west coast of Vancouver Island. Strong westerly winds at times reaching gale force will abruptly flow into Juan de Fuca Strait and become strong southerlies as they enter the southern end of the Strait of Georgia.

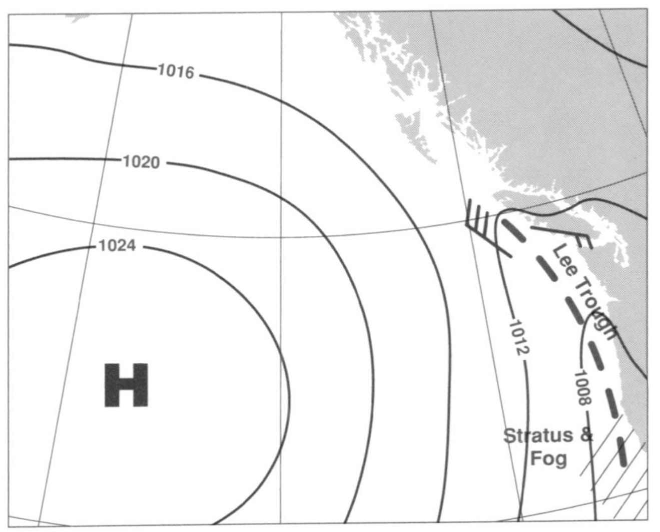

Strong winds occur to the west of the lee trough along the west coast of Vancouver Island with light winds near the coast. (Pressure values in millibars)

Summer Storms

c) Stratus surge

When a lee trough exists along the west coast of Vancouver Island with light winds extending well out from the island a potentially dangerous phenomenon can take place. This event is sometimes called an "alongshore **stratus surge.**"

Fog and low stratus which has been lying along the northern California coast begins to slowly move northward. This northward motion can take days and sometimes stalls temporarily along the Washington coast. However, with a very minor change in the pressure pattern it begins to surge rapidly northward along the west coast of Vancouver Island. Usually it stops near the northern end of the Island but occasionally it moves right up into Hecate Strait. As the stratus and fog rushes northward, the winds abruptly change from light easterlies to strong southerlies. These strong winds often rise to gale force as they round the Brooks Peninsula and storm-force winds are not unknown. Local fishermen call these winds, **fog winds.** The air temperature will drop sharply with the arrival of the fog.

There are no reliable indicators to warn the mariner of this stratus surge as the northward movement of the stratus from the Washington coast is so rapid. The mariner off the west coast of Vancouver Island during the light winds of summer should just keep a close weather eye on the approach of fog from the south and listen for any warnings or changes to the marine forecast.

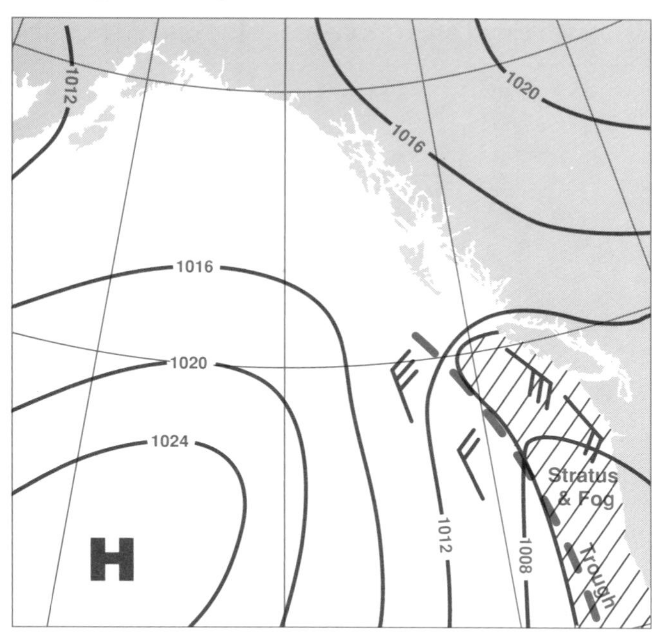

A stratus surge along the west coast of Vancouver Island brings strong southerly winds, stratus clouds and fog. (Pressure values in millibars)

Thunderstorms

Air mass thunderstorms occur most often during the winter in the very cold air that follows the passage of a cold front. They often are spread over a large expanse of the northeastern Pacific and frequently form into **squall lines** which move toward the coast about 12 to 24 hours after the frontal passage.

Occasionally during the summer, a thunderstorm cloud cell develops over land during the late afternoon or evening and drifts over the water. This occurs most frequently over the southern sections of the Strait of Georgia. These are often individual cells which can give very intense thundershowers accompanied by hail and lightning. They are usually short lived. It is often possible to avoid these storms by simply altering course.

Frontal thunderstorms, however, form a more or less continuous line of activity along a front, allowing little opportunity to avoid the storm by changing course. In addition, frontal thunderstorms are often embedded in other clouds and the precise area of thunderstorm activity is difficult to separate from the general cloud and rain accompanying the front.

Initially, when a thunderstorm cloud, or cumulonimbus, is growing, it consists of a central core of upward moving air. Winds around the cloud, though not strong at first, blow in towards the centre. As the storm cloud reaches its maximum intensity, a strong downdraft of cooler air develops. Rain begins and is accompanied by lightning and thunder. Strong, gusty winds blow down from under the cloud and spread outwards. A dark menacing roll or arch cloud often forms at the forward edge of the thunderstorm. The strongest winds occur ahead of the storm, blowing in the direction of the storm's motion.

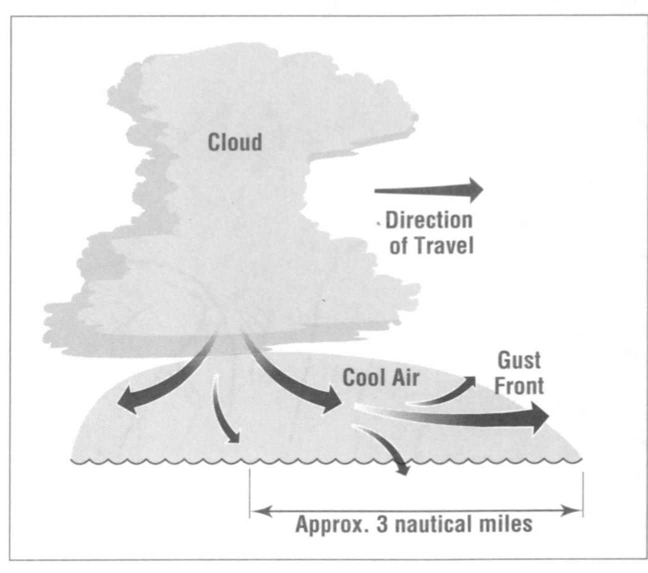

The advance of the cold air is much like a miniature front. The zone of gusty winds may extend two to three miles in advance of the cloud and precipitation area. Wind gusts as high as 50 knots are not uncommon. Since the winds originate in the cloud they blow downwards and can be of concern to sailing craft.

Thunderstorms

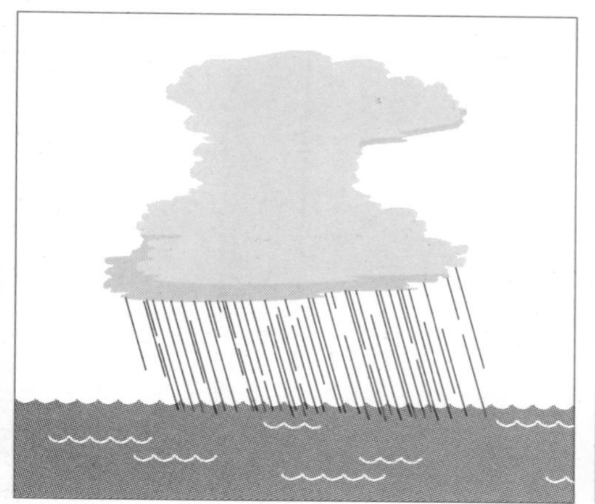

The heaviest rain occurs directly under the core of the cloud and results in poor visibility. Heavy rain will last from 5 to 15 minutes and then decrease as the cloud begins to dissipate. The lifespan of the thunderstorm is usually less than an hour.

"The Pride of Baltimore", a replica of a nineteenth century clipper ship, was caught by a downdraft from a cumulonimbus cloud in May 1986, in the West Indies. The Pride of Baltimore was knocked down and on its way to the bottom in less than two minutes.

Another phenomenon associated with thunderstorms is the **waterspout.** It is a rotating funnel of cloud that extends down from the base of a thunderstorm cloud. The first sign that a waterspout may form is when the cloud sags down in one area. If this bulge continues downward to the sea surface, forming a vortex beneath it, sea water will be carried aloft in the lower twenty or thirty metres. The average diameter of a waterspout is from seven to 20 metres although some exceptional systems may reach 100 metres across. Waterspouts look like tornadoes, but are not as severe and usually do not last more than 10 or 15 minutes. Although some immature waterspouts are very small, they should definitely be avoided as they can change to more violent systems without warning.

CHAPTER THREE

Wind

Introduction

The wind has been of interest to mankind down through the ages, and in all parts of the world. It has propelled sailing vessels, turned windmills and brought the rains to water the fields and forests. It has also brought destruction and death through the awesome strength of its fury. Even though the wind is always with us, it stills retains a certain degree of mystery and respect.

The overall patterns of the winds are quite well understood. It is the daily variation of the winds, where they blow and how strong, that remains a constant problem for meteorologists to forecast.

The problem of forecasting the winds becomes even more difficult as the rugged coastline of British Columbia modifies the wind flow pattern in many different ways. These localized effects cause some areas to have generally light winds and other areas to have frequent and dangerously strong winds. The difference between these two situations often depends on the specific shape of the nearby coastal topography.

If mariners can understand why the winds are much stronger in some areas during certain weather patterns, they can more effectively use the marine forecasts. Mariners will also be better able to interpret the observations available from the weather reporting stations, which reflect the influence of the neighbouring topography. Under some conditions the local weather observations may not be representative of the surrounding marine area.

This symbol for strong winds will be used in chapter six to identify locations where strong winds may result in hazardous conditions.

Pressure

Wind is simply air in motion. Air is set in motion to compensate for differences in barometric pressure between different areas. Whether on the grand scale of the entire earth or on the smaller scale of a bay or inlet, wind results from air moving from an area of higher pressure to one of lower pressure.

The greater the difference in pressure between two points (this is called the pressure gradient) the stronger the winds. On a weather map, lines called "isobars" are drawn between points of equal pressure. When isobars are closely spaced, the pressure gradient is strong and thus the winds are also strong. This is comparable to topographical charts which indicate a steep slope in the terrain by the closely spaced contours.

If the earth did not rotate, the wind would blow directly "downhill" from high to low pressure. However, because the earth does rotate, the wind turns to the right (in the northern hemisphere) and blows counterclockwise around the low—more or less parallel to the isobars—and clockwise around the high.

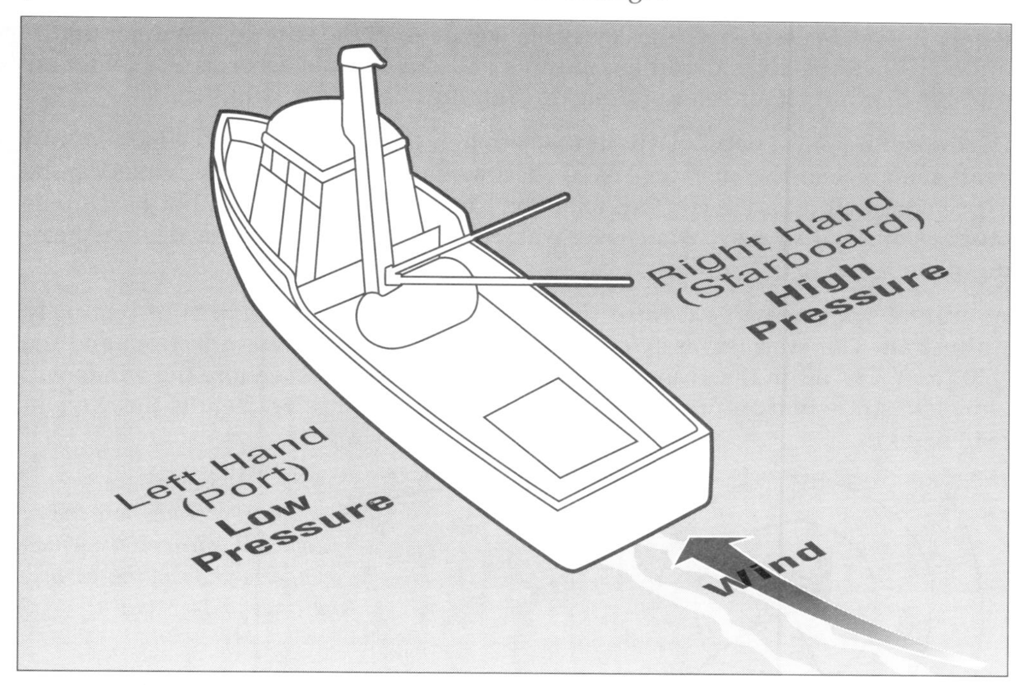

If the wind is blowing at your back, the lower pressure is on your left hand side and higher pressure in on your right.

Stability

The **stability** of the air over open water is largely determined by the temperature difference between the water and the air.

When warm air moves over colder water, the water cools the air near the surface. This sets up a pattern which is called **stable** for the colder, heavier air is below and the warmer, and hence lighter air, is above. There is no tendency for either the warm or cold air to move to change this pattern. However, when cold air moves over warmer water the lowest layer of the air is heated up. This warmer air, which is lighter than the cold air, then begins to rise so that it is above the cold air. The air in this situation is **unstable** and commonly occurs in winter after the passage of a cold front.

What does stability have to do with the wind?

The winds aloft are generally stronger than surface winds because friction slows down the air moving close to the surface. When the air is stable, the "frictionless" winds aloft slide easily over the cooler blanket of air at the surface, leaving it undisturbed. When the air is unstable, updrafts and downdrafts occur as the colder air replaces the warmer air below. Strong downdrafts cause gusty winds.

During sunny days, particularly in the summer, the land is heated, which in turn heats the air near the surface. The air becomes unstable and gusty winds can be expected on land and very close to shore. The sea, however, does not heat up as much as the land so gusty winds over water are not usually caused by daytime heating alone.

As frontal systems move toward the coast they often bring an area of relatively warmer air. This warmer air is found in the sector between the warm front and the cold front. The air in the "warm sector" is more stable and therefore the winds will blow less strongly than in the cold air, even if the pressure gradient is the same in both regions.

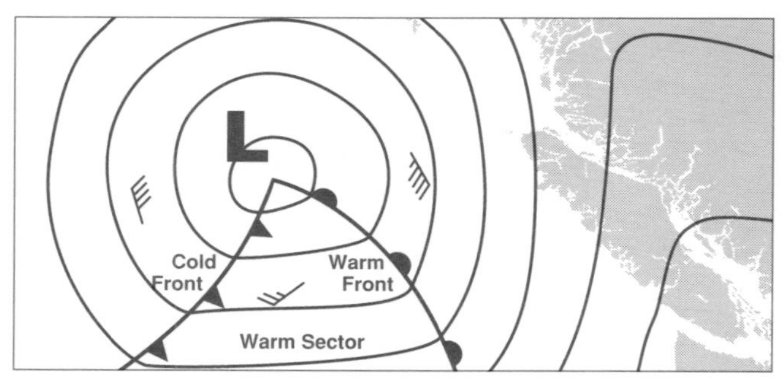

When the isobars are equally spaced in both the cold and warm air the winds will be lighter in the warm air due to its stability.

Lee effects

When the winds blow against a steep shoreline bluff or over rugged terrain onto the water surface, gusty turbulent winds result. Eddies often form downwind of the cliff face which create stationary zones of stronger and lighter winds. The zones of strong winds are fairly predictable and usually remain stationary as long as the wind direction and stability of the airstream do not change. The lighter winds, which occur in areas called wind shadows, can vary in speed and direction, particularly downwind of higher cliffs.

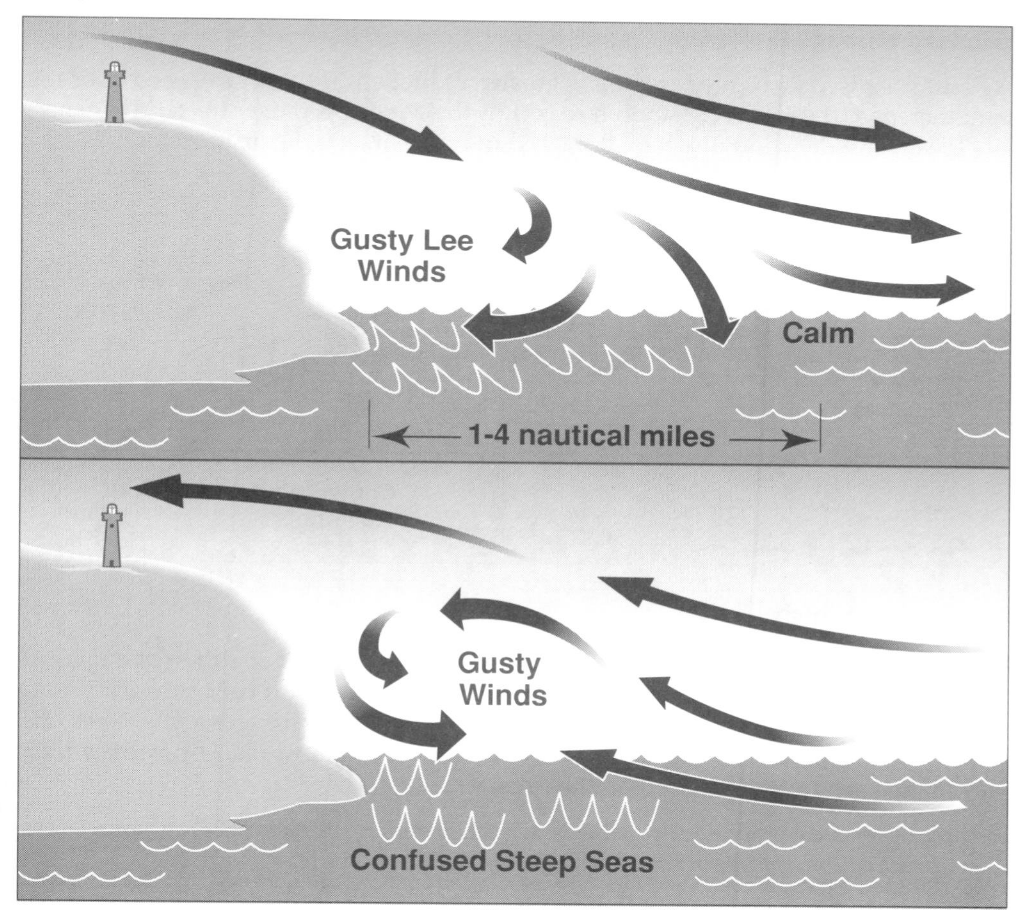

Above: *Beneath the cliffs the wind is usually gusty and the wind direction is often completely opposite to the wind blowing over the top of the cliff.*
Below: *Smaller but reversed eddies may be encountered in onshore winds near cliffs.*

Friction effects

The winds that blow well above the surface of the earth are not strongly influenced by the presence of the earth itself. Closer to the earth, however, frictional effects decrease the speed of the air movement and change the direction slightly.

As the wind speeds are reduced by friction they turn more sharply toward lower pressure. This means that if the wind is at your back, the winds turn counterclockwise (ie. they are **backed**) and blow more directly toward the lower pressure. A southerly wind, for example, becomes more southeasterly when blowing over rougher ground.

As our coast is very rugged there is a greater reduction in the wind speed over the land than over the relatively smooth sea. Thus the wind speed over the land is generally lower, and the wind direction is backed more, than over the open water.

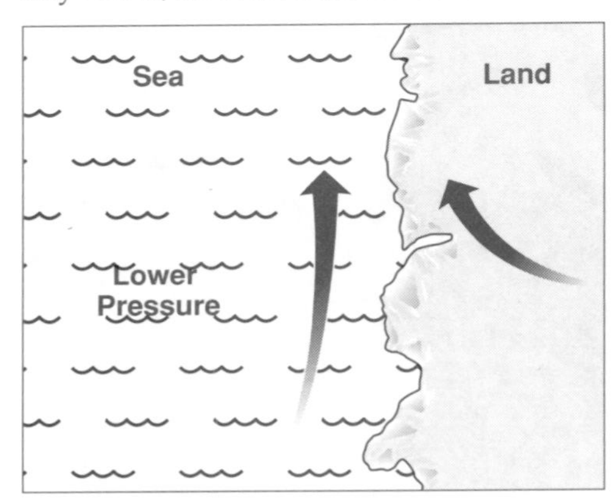

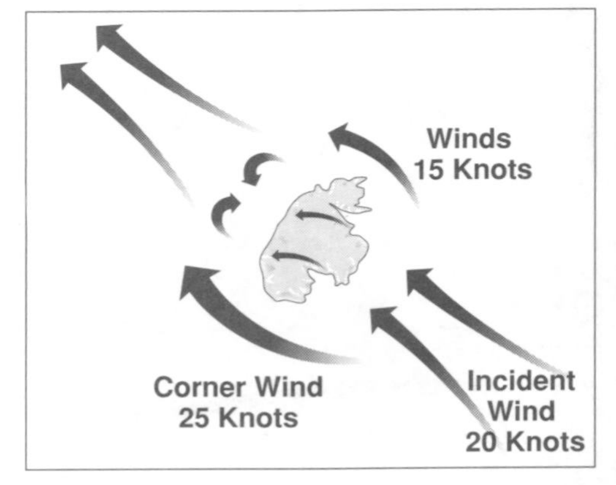

a) Corner winds

When the wind is at your back and the coast is on your right, the different angles of the surface winds over land and water cause the airstreams to converge. This **convergence** creates a band of wind which is about 25 percent stronger a few miles offshore. The convergence of airstreams causes an increased flow of air upwards which may in turn produce more cloud in this area.

In the opposite case, when the wind is at your back and the coast in on your left, a **divergence** of the airstream results in a band of lighter winds.

These effects are seen when the wind blows past an island or around a headland. Turbulent winds are often found to the lee of the island or headland.

Meteorologists refer to these phenomena as **corner winds.**

Friction effects

The Brooks Peninsula, which stands at right angles to the overall orientation of the west coast of Vancouver Island provides a good example of local wind modification due to corner effects. Winds at Solander Island, just off Cape Cook, are invariably much stronger than those reported from either side of the peninsula. The area is also noted for its dangerous sea conditions.

The strongest winds are usually the southeasterlies, just ahead of a cold front. Here the air is relatively warm and stable. When the air is stable it is less able to rise up over the peninsula but flows around it instead. Unstable air, however, will flow more readily over the peninsula and hence does not cause such a marked corner effect.

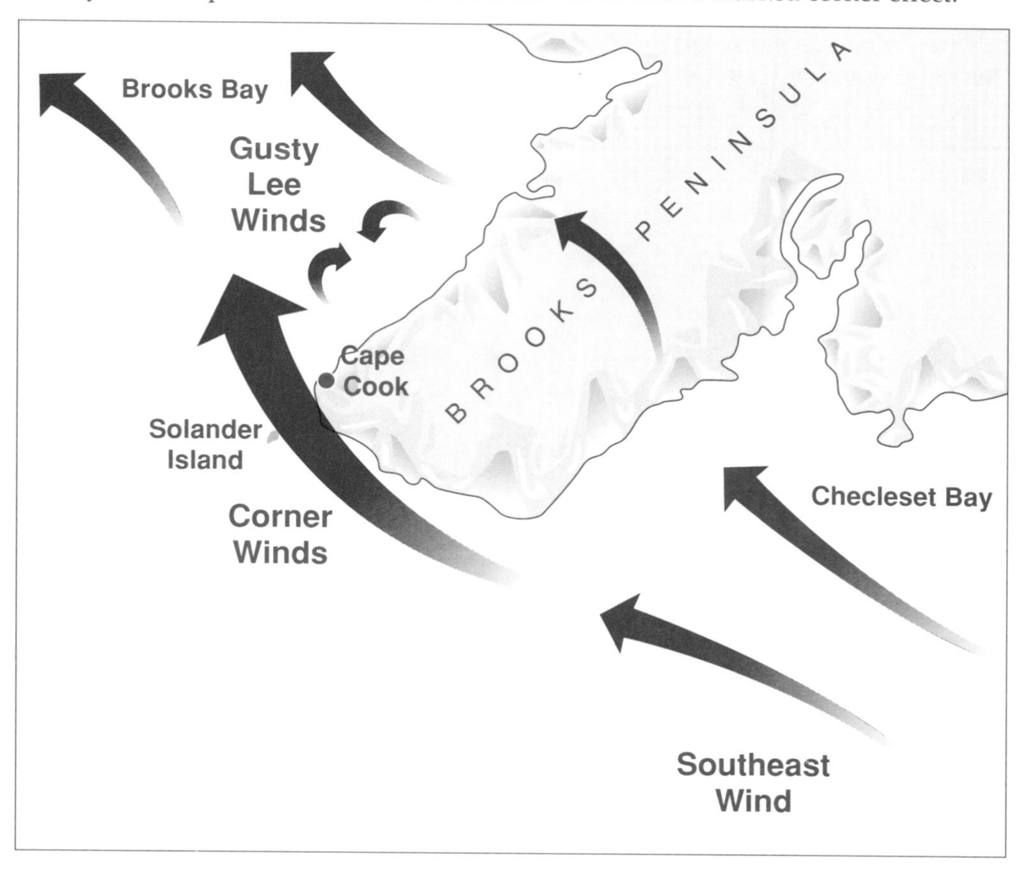

Winds flow around and over Brooks Peninsula causing very strong winds off Cape Cook with gusty, lee winds in Brooks Bay. Solander Island reports give a good indication of corner wind speeds.

Friction effects

b) Offshore wind maximum

The convergence of the winds caused by frictional effects occurs every time that a storm approaches the B.C. coast. As the storm moves closer to the coast the mountain ranges of Vancouver Island and the Queen Charlotte Islands steer the winds into the southeast and reduce their speed.

The reduction of the speed has been verified by reports which indicate that the coastal reporting stations often have lighter winds than those further offshore.

As coastal winds are backed into the southeast they converge with the more southerly winds offshore and result in a band of stronger winds and higher seas. The location of this band of strong winds varies somewhat with the strength and direction of winds blowing onto the coast but generally it is between 3 and 15 miles offshore.

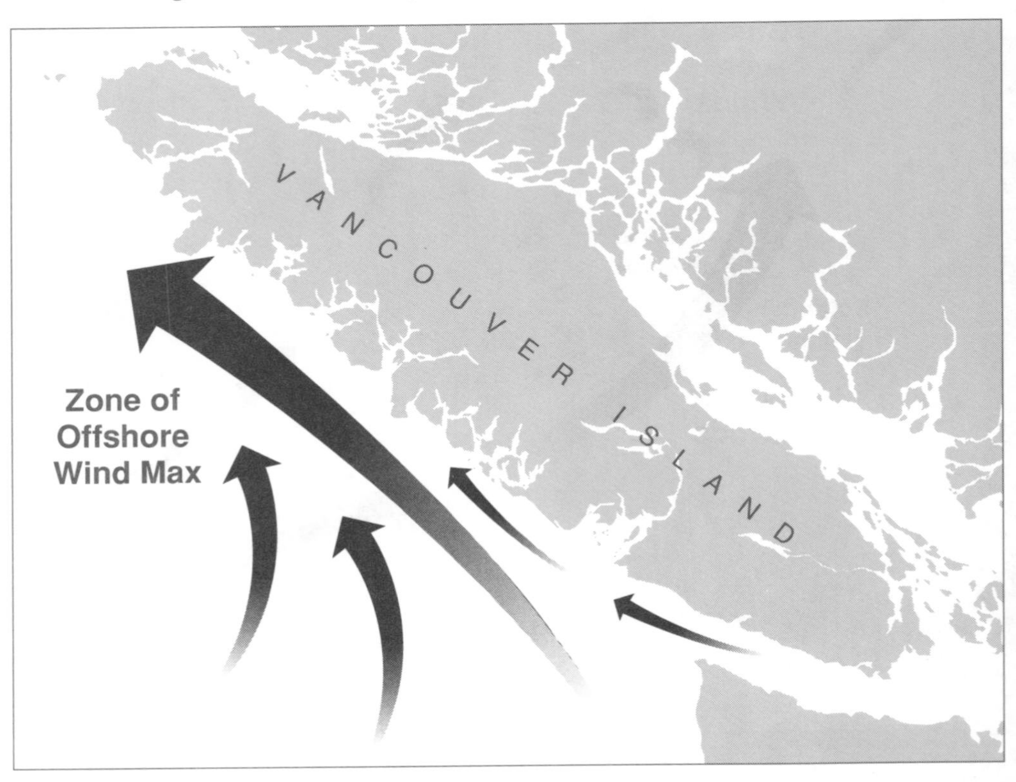

Zone of offshore wind maximum

Gap winds

When winds are forced to flow through a narrow opening or gap, such as through an inlet or between two islands, the wind speed will increase and may even double in strength. This effect, called funnelling, is similar to pinching a water hose to create higher speeds.

The coastal topography can also change the direction of the wind by forcing it to flow along the direction of a pass or through a strait. This is referred to as channelling.

When winds have been modified by both funnelling and channelling they are called **Gap winds.**

The constricted channels not only increase wind speeds but also strengthen tidal currents. Care is necessary when the increased winds are directed against strong tidal currents. The resulting steep, breaking waves may produce much rougher waters than in more open areas.

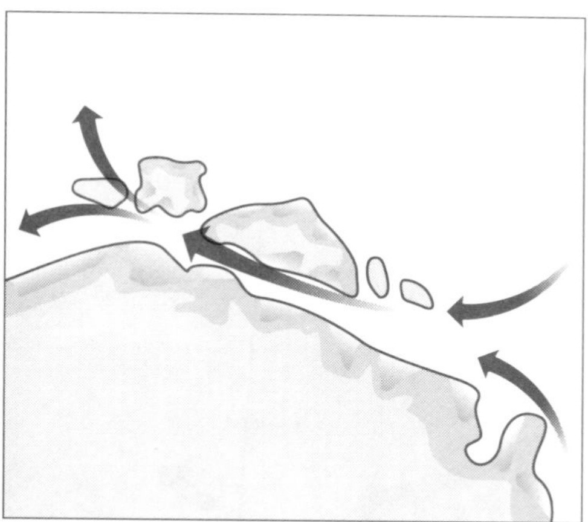

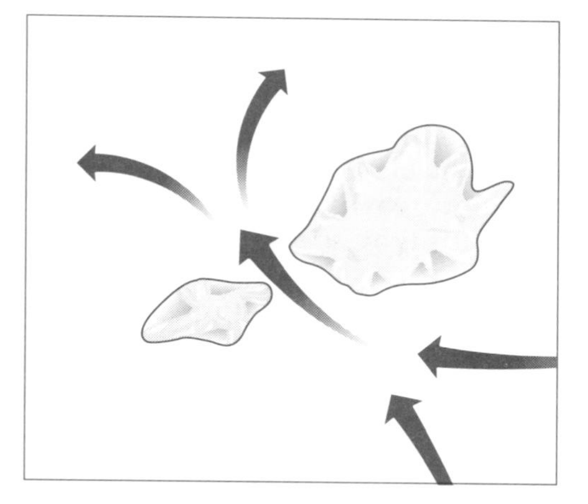

Strengthened winds due to the constriction of the flow of air caused by two land masses.

Gap winds

An example of gap winds occurs in Juan de Fuca Strait.

When there is a ridge of high pressure off the coast, northwesterly winds will blow along the west side of Vancouver Island. These winds will be forced through Juan de Fuca Strait as westerlies and are often increased in strength, sometimes reaching gale force.

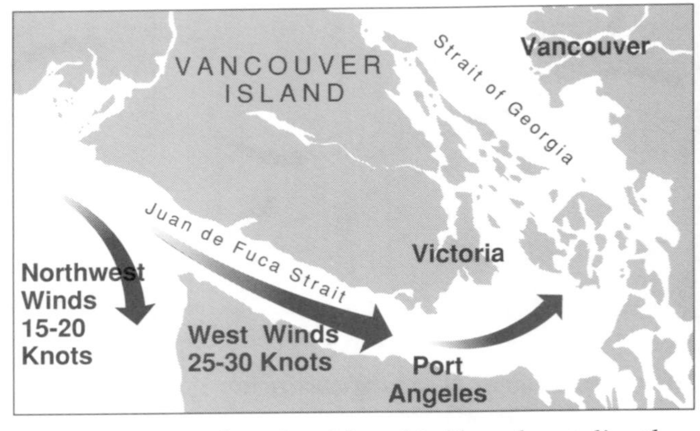

Winds are strengthened and forced to blow almost directly along the axis of Juan de Fuca Strait.

When a low pressure area or a front approaches southern B.C., the winds are generally southeasterly along the outer coast. Both Pachena Point and Carmanah Point will report southeasterlies in this case. The winds in Juan de Fuca Strait, however, are easterly due to channelling. As the air in advance of the front is usually stable, the winds are more easily channelled. Light easterly winds near Victoria will gradually increase along the Strait to reach 25 to 30 knots at the western entrance.

Onshore winds are often blocked by the mountain ranges along the B.C. coast. However, where passes occur, strong winds blow through the opening and down the other side. The winds increase through the constriction and result in a strong flow on the lee side of the mountain range.

A good example of this is found on the east side of Vancouver Island. The mountain ridge forming the backbone of Vancouver Island has a well-marked opening from Barkley Sound past Port Alberni to Qualicum Beach. The wind blows in a strong southwesterly flow along the Alberni Inlet, through the pass in the middle of Vancouver Island, and continues down over the east coast of the Island to the Strait of Georgia. These winds, known locally as "Qualicums", usually spring up during the afternoon on hot summer days.

At night and during the early morning when the air is often stable, the wind flowing through the pass does not drop down to the ground on the east side of Vancouver Island but remains aloft. However, with daytime heating, turbulence mixes the air and brings the strong winds down to the surface. Conditions such as these can be quite treacherous for the unwary mariner.

During the morning the weather can be clear with glassy seas on the Strait of Georgia but, suddenly in the afternoon, southwesterly winds hit the Qualicum Bay area with speeds up to 40 knots. The zone of strong winds is about 2 miles wide, often reaching across the Strait of Georgia to the northern tip of Lasqueti Island. Here the winds diminish and turn southward into Sabine Channel or along the west coast of Lasqueti Island. Sisters Island is the only local observation site that reports the "Qualicum."

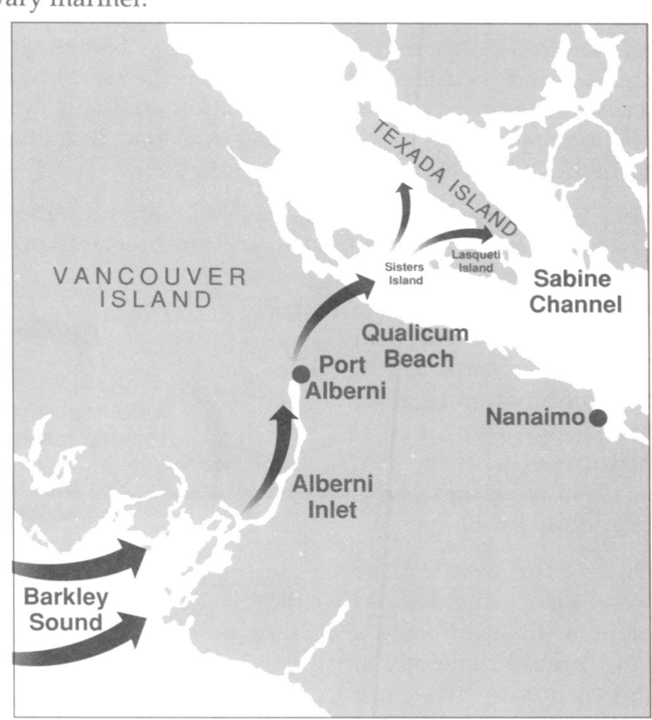

Land and Sea breezes

The local effects discussed so far have been primarily due to the deflection or blocking of the wind by adjoining land areas in much the same way as a river speeds up or turns in response to the shape of its banks. The following effects, however, depend on temperature differences between adjoining regions.

Land and Sea Breezes are only observed when the prevailing winds are light and when strong daytime heating occurs. This is often the case in the summer when a large area of high pressure dominates the weather pattern.

The **sea breeze** blows from sea to land and occurs when the air over the land is heated more rapidly than the air over the adjacent water surface. As a result, the warmer air rises and the relatively cool air from the sea flows onshore to replace it. As the day progresses the sea breeze circulation gradually strengthens and extends further offshore. Speeds of 10 to 15 knots can extend 15 miles out to sea by late afternoon. The wind will generally veer as it strengthens.

The irregularity of our coastline can add many complications to this simple pattern of sea breezes. For example, when the sea breeze is funnelled through an inlet, the sea breeze can become a wind of up to 30 knots. In Juan de Fuca Strait and Queen Charlotte Strait, the combination of a sea breeze with an overall inflow situation can result in afternoon winds of up to 40 knots.

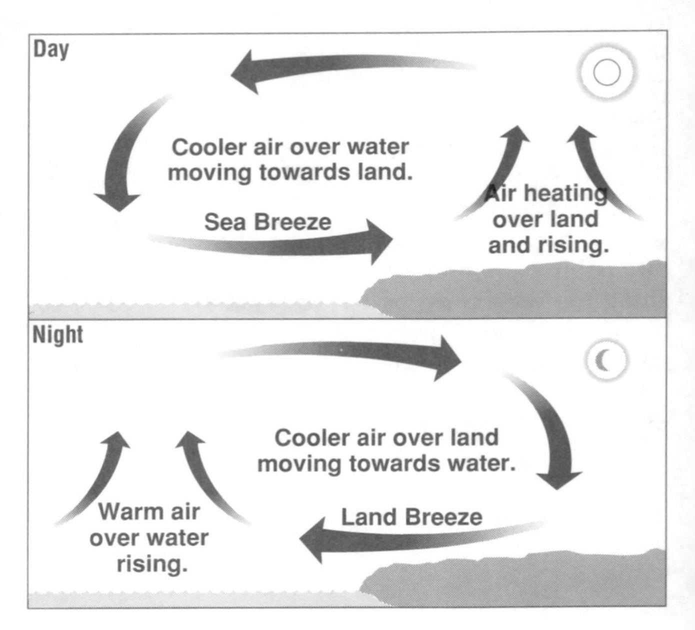

During the evening the sea breeze subsides. At night, as the land cools, a **land breeze** develops in the opposite direction and flows from the land out over the water. It is generally not as strong as the sea breeze, but can be quite gusty.

Katabatic and Anabatic winds

Winds in valleys and along mountain slopes behave in a very similar way to land and sea breezes.

During the day the sides of the valleys become warmer than the valley bottoms since they are more exposed to the sun. As a result, the winds blow up the slopes. These daytime, upslope winds are called **anabatic winds.** Gently sloped valley sides, expecially those facing south, are more efficiently heated than those of a steep, narrow inlet. As a result, valley breezes will be stronger in the wider valleys.

At night, the air cools over the mountain slopes and sinks to the valley floor. If the valley floor slopes down to the coast, as it does at the head of most inlets, then the cold air flows down the slope and out to sea. The cool night winds are called **drainage winds** or **katabatic winds** and are often quite gusty and usually stronger than the daytime anabatic winds.

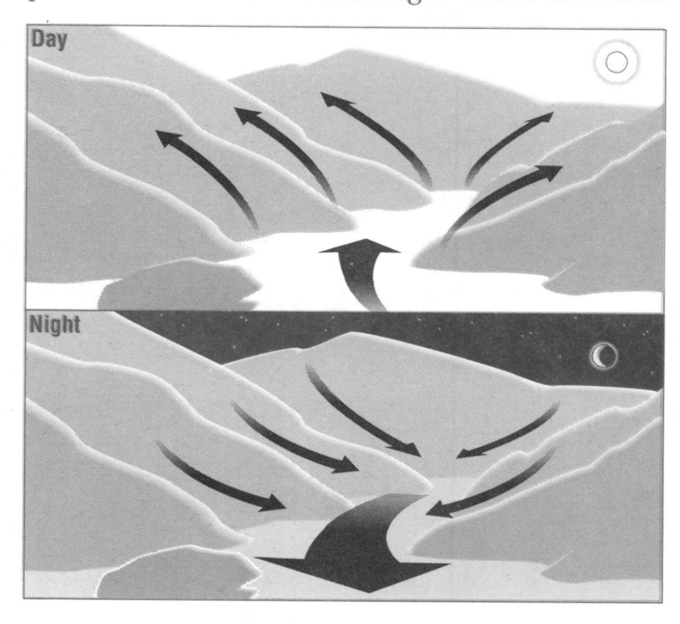

After preparing for an overnight anchorage in a calm secluded cove, the onset of gusty downslope winds from a valley at the head of the cove can be an unpleasant surprise.

These mountain valley winds can contribute to the land and sea breezes and result in increased wind speeds.

STRONG WINDS AND HIGH SEAS

CROSS WAVES

STEEP WAVES

a) Wave–Current Interactions
b) Rips and Overfalls
c) Shoaling and Refraction
d) Fraser River Bar

CHAPTER FOUR

Sea State

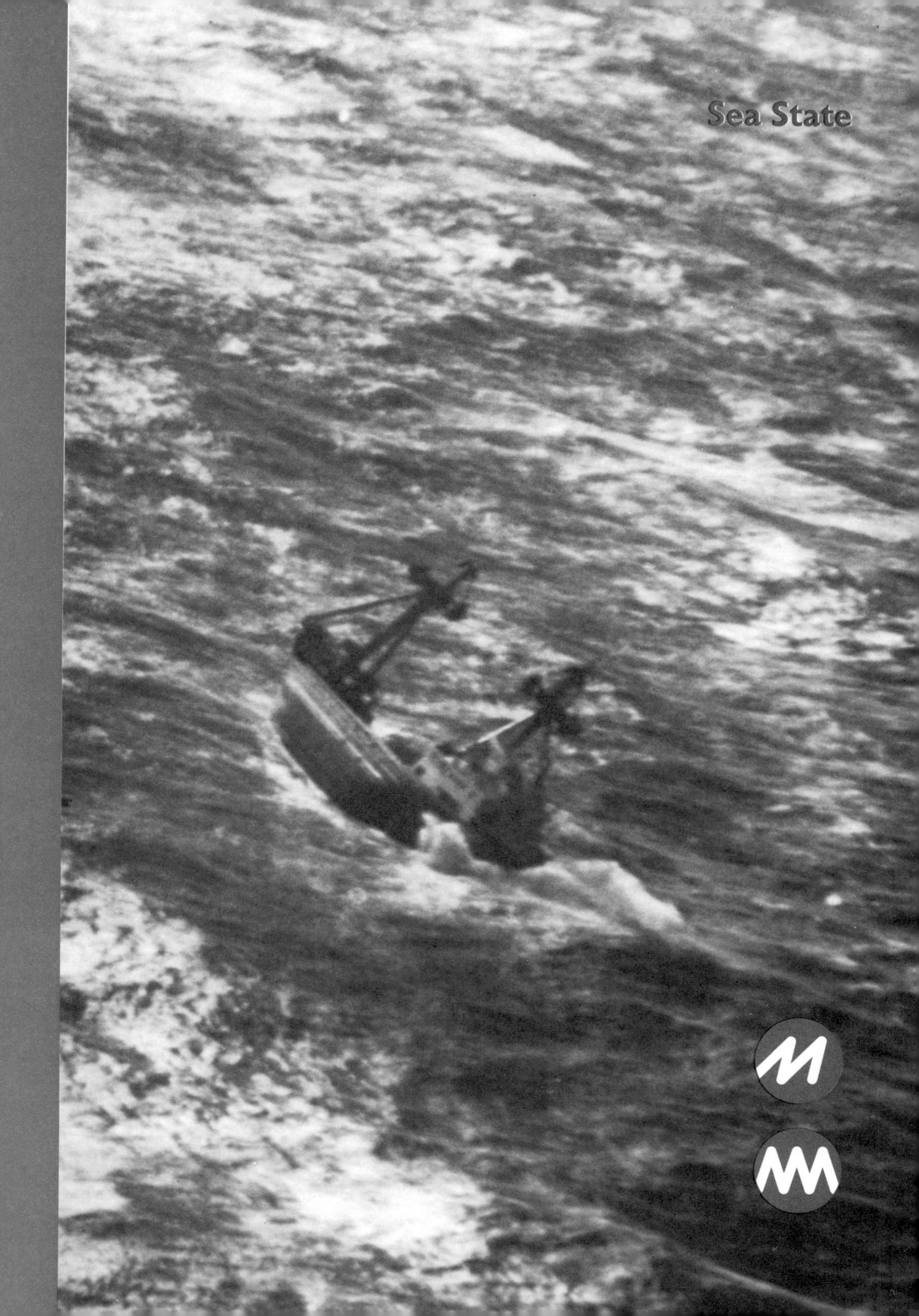

SEA STATE

Strong Winds and High Seas

This symbol for high waves will be used in chapter six to identify locations where large waves, formed by wind alone, may result in hazardous conditions.

The most hazardous sea conditions along an exposed coast result from the waves generated by storm and hurricane force winds. Generally, the height of waves depends upon the wind speed, which, in turn, is related to the central pressure of the low. A 950 millibar low typically has stronger winds than a 990 millibar low.

In addition to wind speed, wave height depends on two other factors:

1. **Fetch—The distance which the wind blows across the water from the same direction and with constant speed.**
2. **Duration—The length of time the wind persists without changing direction or speed.**

The near-shore waters have definite fetch limitations whenever the winds blow off the coast. The fetch of the wind is also limited by a frontal system as the winds usually change direction after the frontal passage.

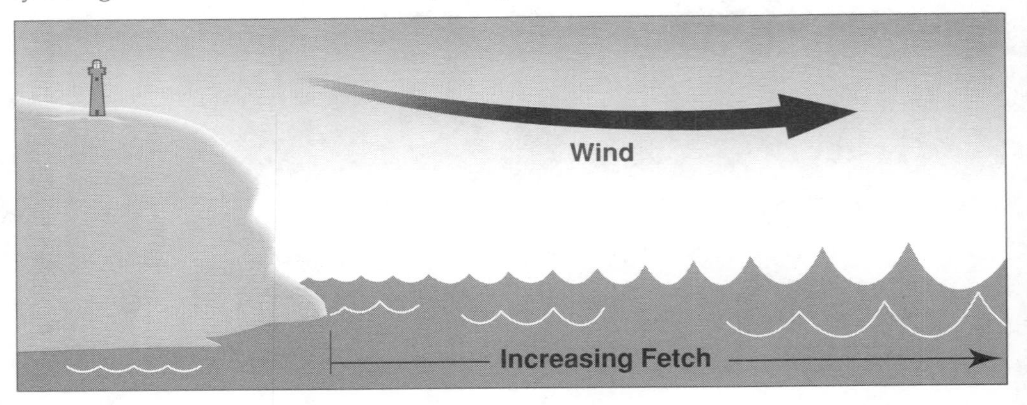

Duration is linked to the speed of advance of the weather system. For example, coastal lows advance rapidly onto the coast with frontal movements of 35 knots or more. Although storm force winds often occur in a winter storm, their duration may be fairly short, thus limiting the development of high seas. Conversely, a slower moving system can have gale force winds blowing from the same direction over a longer period of time. The greater duration can offset the lower wind speeds to give wave heights that are similar to those associated with more intense coastal lows.

SEA STATE

Strong Winds and High Seas

The phrase "**sea state**" as used in this manual and consistent with Canadian marine forecasts and warnings, refers to the significant height of the combined wind wave and swell. **The significant wave height is the average height of the highest one-third of all the waves present.** The significant wave height approximates what an observer at sea would likely report.

When using the sea state values in the marine forecast, keep in mind that:

- the forecast value is the significant height of the **combined wind wave and swell in metres**
- there will be waves present that are half the forecast value
- **the maximum individual wave over 3 or 4 hours can be double the forecast value**

Below is a general guide to the relationship between the strength of wind, the depth of the associated low and the resultant sea state.

Term	Wind Speed	Approximate Central Low Pressure Value	Fully Developed Seas in Exposed Locations
Strong winds or small craft warning	20–33 knots	985–995 mb	2–5 metres
Gales	34–47 knots	975–990 mb	5–8 metres
Storm Force	48–63 knots	960–980 mb	8–12 metres
Hurricane Force	64 knots or greater	Lower than 960 mb	Greater than 12 metres

The sea state figures in this table assume that the winds blow over open water from the same direction for 12 to 24 hours. If the duration of the winds is less than 12 hours or the fetch is limited then the sea state values will not be as high. Likewise if the winds blow for longer than 24 hours with unlimited fetch, then the values could be higher.

Is every seventh wave the biggest? Here's what B. Kinsman, an oceanographer, has to say: "As I have traveled about, I have found every integer from three to nine enshrined in the folklore. If you put your faith in any 'pet integer' and have the temerity to prove your faith in a sailing dinghy, you will sooner, rather than later, be slapped silly by a wave numbered 'pet integer plus one'." The answer is NO!

Strong Winds and High Seas

In areas such as Hecate Strait, Dixon Entrance, and the Central Coast, seas will be less severe because of **sheltering** by adjacent landforms for some wind directions.

For many storms the northern end of Hecate Strait is sheltered from developing high seas. However, when southeasterly gale-to-storm-force winds blow over a fetch distance of about 300 miles seas can build to 8 to 9 metres over the shallow waters near Bonilla Island. This usually occurs two or three times every winter.

In many situations **wave steepness** becomes more critical than wave height. A boat may be able to ride a long, high wave by climbing up one side and sliding down the other. However, in cases of steep waves travelling close to one another, serious trouble results when a boat's stern gets hung on one wave crest while the bow is driven under the next one.

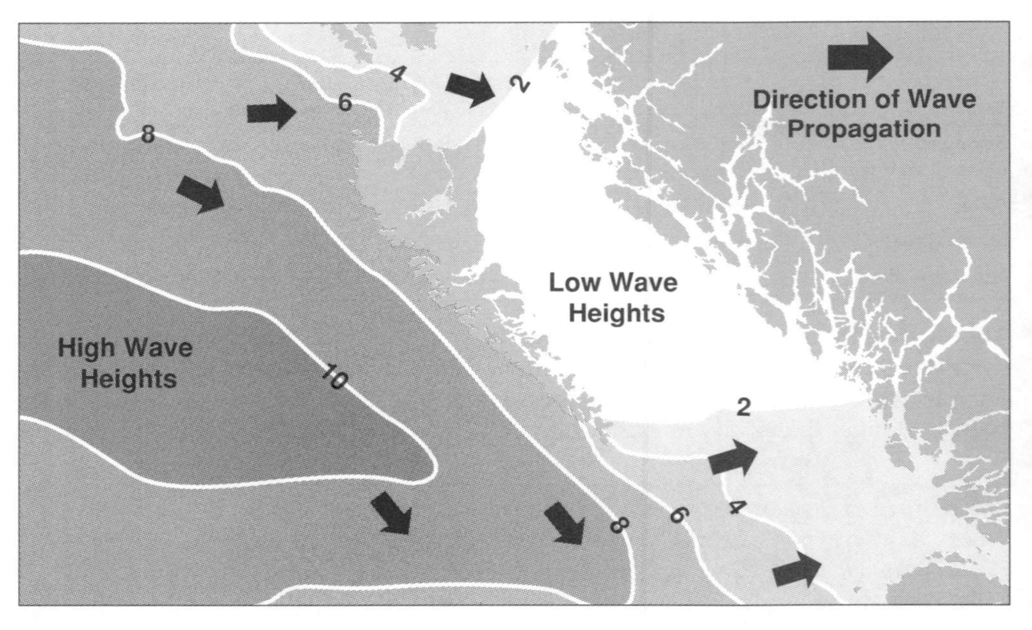

Wave height field (in metres) for Northwest gales onto the west coast of the Charlottes. Hecate Strait is sheltered from high wave heights due to limited fetch.

Strong Winds and High Seas

High seas of any kind are dangerous if you are not prepared. You should take special care in the following situations.

*In **beam seas** excessive roll can cause cargo to shift, creating a dangerous list. This could cause the vessel to capsize. Strong breaking waves could also capsize the vessel.*

*In **following seas**, a vessel may lose stability on a wave crest. If the vessel is overtaken by a wave crest, broaching may occur.*

*In **quartering seas**, the problems of beam and following seas are combined. Quartering seas represent the most dangerous situation in severe weather.*

Strong Winds and High Seas

Typical sea states over the open ocean for various wind conditions.

Strong Winds. (near 20 knots)
Waves take on a more pronounced long form with many white horses.

Strong Winds. (near 30 knots)
Sea heaps up and white foam from breaking waves begins to blow in streaks along the direction of the wind.

Strong Winds and High Seas

Gale Force Winds. (near 40 knots)
Foam streaks become very dense in well-marked streaks.

Storm Force Winds. (near 50 knots)
Very high waves. Sea has a white appearance from the great patches of dense foam streaks. Visibility affected.

Cross Waves

This symbol for steep waves will be used in chapter six to identify locations where waves, which result from the interaction of two or more waves, may result in hazardous conditions. This includes wave–current interactions.

Cross waves refers to the condition that occurs when one train of waves is moving at an angle to a second group of waves. The seas develop pyramidal shapes with short, sharp, wave crests and appear confused. Depending on how big the wave heights are in each wave train, cross-wave conditions range from uncomfortable to hazardous for smaller vessels. When the waves interact with an underlying tidal current to steepen and break, the sea becomes especially confused and shock-like.

Wave crossing occurs in two ways. First, when wind waves and swell are propagating in different directions the waves can cross and interact. This most commonly happens when new waves are being generated over an old swell field, which was produced by an earlier or distant storm. Secondly, wave crossing arises when a front associated with a Pacific storm moves over an area. A marked shift in the wind direction usually occurs with the passage of the front. The wind waves formed with the southeasterly winds ahead of the front will then merge with the westerly waves that arrive after the front.

A (Winter)
Alaska Current
Davidson Current
Confused & Variable Currents
B (Summer)
Alaska Current
Confused & Variable Currents
California Current

Prevailing large-scale surface currents off the British Columbia coast in winter and summer. Bold arrows indicate the major currents which are between .5 and 1.5 knots.

Steep Waves

a) Wave – Current Interactions

Near the coast, waves increase in height, steepen and break as they run onto an opposing tidal or ocean currents. When the current is strong or the waves are large due to high winds, the breaking may be vigorous. These conditions can be quite hazardous depending upon the vessel size and type. Cape Mudge, Nawhitti Bar and Scott Channel are well known areas of strong, wave - current interactions.

Waves moving against a current actually receive energy from the current. The wave length decreases, while the wave height increases. This results in a rapid steepening which may lead to breaking. For example, a 3 metre wind wave would almost double in height on a 5 knot current and would steepen to the breaking point.

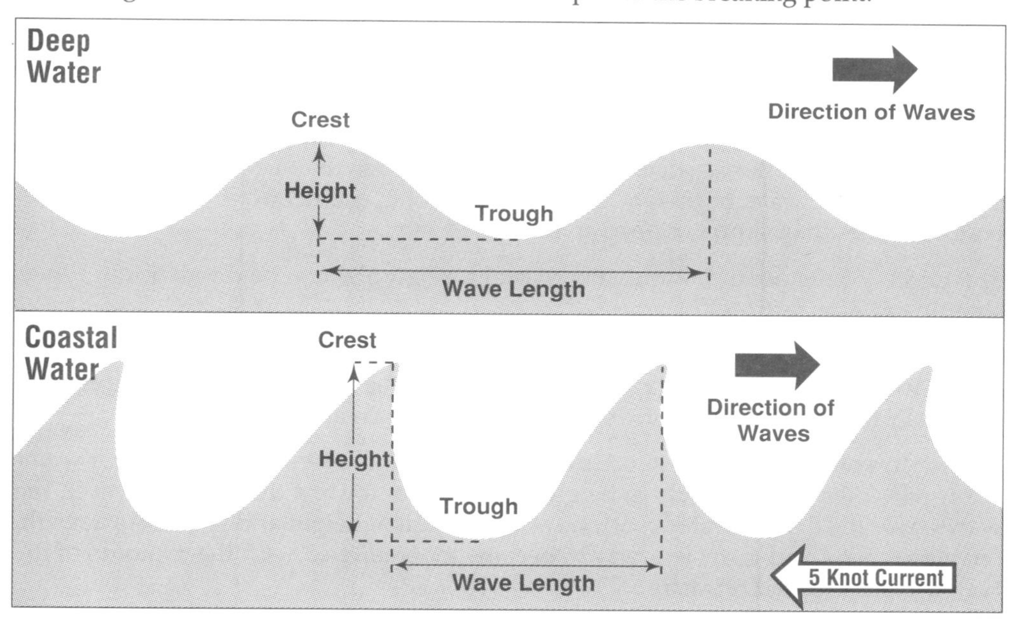

Mariners are advised to plan routes that take advantage of the stage of tide (ebb-flood) and to use the weather forecast along with local observations of wind and sea to avoid dangerous spots. ***Sailing Directions*** *can be of assistance in determining these areas.*

Steep Waves

b) Rips and Overfalls

Rips are a turbulent agitation of the water generally caused by the interaction of currents and wind waves. In shallow waters the irregularity of the sea bottom can also create these short breaking waves.

When the current is predominantly tidal, the rip is often referred to as a **tidal rip**. Rips frequently occur in the entrance to a pass or narrows such as Active Pass, Porlier Pass and Johnstone Strait.

Overfalls are areas of turbulent water caused by strong currents setting over submerged ridges or shoals. A severe overfall can produce a sharp drop in the water level. Overfalls often occur on the north side of Rose Spit near the transition from shallow to deep water where the tidal current is strong with speeds up to 3.5 knots.

c) Shoaling and Refraction

The distance between two wave crests, or between two wave troughs, is called the wave length. When the depth of the ocean bottom is less than half the wave length of a wave, then the wave begins to "feel bottom". Long, rolling swell will begin to feel bottom sooner than short choppy waves.

The effects of the ocean bottom and coastal topography can be divided into types: **shoaling and refraction**.

When deep-water waves reach a shoal and start to feel bottom, the waves become higher. The wave crests also move closer together causing the waves to become steeper, until eventually the wave tumbles into breakers or surf. This is called **shoaling**.

When a wave approaches a shoreline at an angle and moves over shallower water the wave slows down, builds in height and turns in toward the land. However, the waves over the deeper water continue at their original height and speed. As a result, the waves bend and grow in a way to become more aligned with the contours of the bottom. This is called **refraction**.

Shoaling affects the height of the waves, but not the direction. **Refraction** affects both. These two effects depend on the change of the wave speed in shallow water. The **Sailing Directions** for your area will give a description of the shape of the local sea bed and unusual shoaling effects.

As a southwest sea approaches Barkley Sound and crosses La Perouse Bank, the waves bend or refract in response to the changing depths and cross in several areas in the approaches. Where the paths of the refracted wave cross, the seas increase in height and steepness. The waves also interact with the tidal currents to produce very confused and hazardous seas at the mouth of Barkley Sound. These affects are also evident at other locations along the British Columbia coast such as off Swiftsure Bank.

d) Fraser River Bar

An example of "wave-current-bathymetry" interaction takes place in the mouth of the Fraser River, off Sand Heads lightstation. Here the ebb tide (a current of about 3 knots is possible) opposes a southwest wind wave moving across Sturgeon Bank and Roberts Bank into the mouth of the river causing much higher and steeper waves. Strong northwest winds and waves also steepen as they oppose the river outflow from the North Arm of the Fraser River.

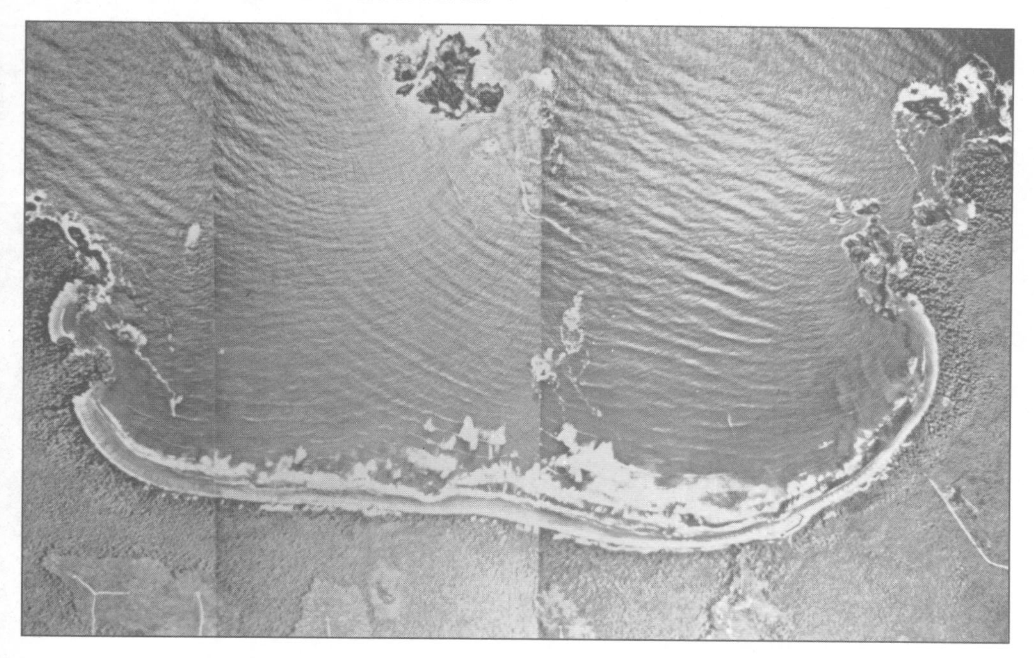

Wave pattern produced by refraction and shoaling around a small island and onto a shoreline.

CLOUDS

RAIN, DRIZZLE AND SHOWERS

FOG

SNOW

ICING

CHAPTER FIVE

Weather

Clouds

Clouds provide one of the keys to understanding the weather. By observing clouds as well as changes in temperature, wind and barometric pressure the mariner can become more aware of weather processes and can perhaps better anticipate forthcoming changes.

Clouds form when the air is cooled. Often the cooling process takes place when the air is "lifted" by winds forcing it up a hill or mountain. A front which is the boundary between two different air masses, one colder than the other, is also important in the formation of clouds. The colder air mass, the leading edge of which is marked by the front, takes on the shape of a dome, much like a puddle of mercury. Air which is forced up over the dome is cooled in the rising, thereby forming clouds.

The diagram below shows the main types of clouds associated with an approaching front. Pictures of these cloud types are on the following two pages.

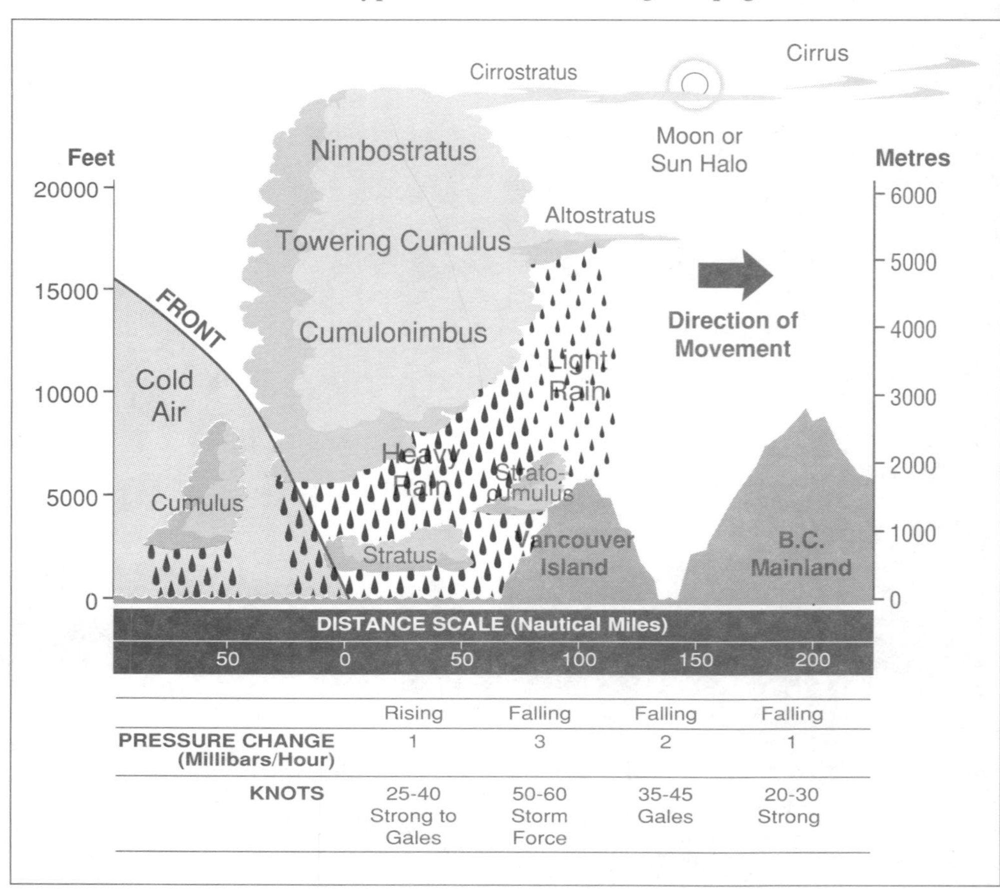

	Rising	Falling	Falling	Falling
PRESSURE CHANGE (Millibars/Hour)	1	3	2	1
KNOTS	25-40 Strong to Gales	50-60 Storm Force	35-45 Gales	20-30 Strong

Clouds

This symbol for weather will be used in chapter six to identify locations where weather such as fog or icing may result in hazardous conditions.

The four main types of clouds, originally named by Luke Howard in 1803, can be combined to include all the cloud types used by meteorologists today.

Cirrus

Cirrus clouds form in the upper realms of the sky and often precede the lower clouds of an approaching storm. The whispy, ethereal nature of cirrus changes into the layered cirrostratus or cirrocumulus as the storm nears.

Stratus

Stratus a low, uniform, featureless layer of cloud, sometimes produces drizzle. When it touches the sea it is called fog. When this layered cloud form occurs in the mid levels of the atmosphere it is called altostratus and at high levels, cirrostratus. Precipitation does not fall from these higher stratus clouds.

Clouds

Cumulus

Cumulus cloud has a flat base, like stratus, but build upward toward the cirrus. Fair weather cumulus have little buildup but can produce showers when they develop into tall towers. If the billowing form occurs in layered clouds it is called stratrocumulus (when at low levels within the atmosphere), altocumulus (mid levels), or cirrocumulus (high levels).

Nimbus

Nimbus is a rain cloud. It occurs in two forms: nimbostratus, which is layered to great heights ahead of a front and produces steady rain, and cumulonimbus which has grown upward from a smaller cumulus cloud and produces heavy rain showers, thunder, lightning and sometimes hail.

Rain, Drizzle and Showers

The marine forecast mentions rain, drizzle or showers when they are expected to significantly restrict visibility. Each of these three forms of precipitation fall from different types of clouds.

Rain is usually associated with a frontal system which spreads an extensive blanket of low, dark looking cloud over the region and often gives several hours of continuous precipitation. Rain often falls from nimbostratus clouds which can extend to a height of 20,000 to 30,000 feet above ground.

Drizzle is a fine, almost mist like precipitation, which often occurs with the passage of a front. However, it can also fall from a thinner layer of stratus clouds found between frontal systems.

Showers differ from rain in that they only occur for short periods of time with breaks in the precipitation. Showers fall from well-developed **cumulus** clouds. The heaviest showers occur with **cumulonimbus** clouds and are often associated with hail, thunder and lightning. Along the west coast the heaviest showers generally occur after a frontal passage with cold northwesterly winds.

A typical summer coastal scene with cumulus and cirrus clouds.

Fog

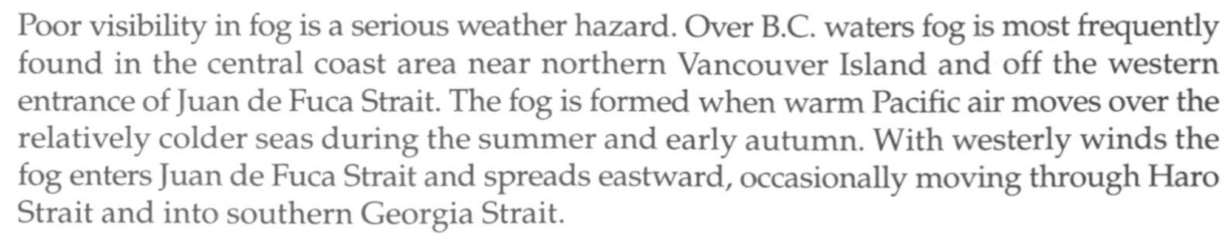

Poor visibility in fog is a serious weather hazard. Over B.C. waters fog is most frequently found in the central coast area near northern Vancouver Island and off the western entrance of Juan de Fuca Strait. The fog is formed when warm Pacific air moves over the relatively colder seas during the summer and early autumn. With westerly winds the fog enters Juan de Fuca Strait and spreads eastward, occasionally moving through Haro Strait and into southern Georgia Strait.

Radiation Fog forms over land during the early morning. It may drift over the water when light breezes blow from land to water during the night. It is primarily a problem in harbours and estuaries. Radiation fog rarely spreads out far to sea and often lifts off the water surface to form low stratus cloud. After the sun comes up, the fog dissipates over land and then clears more slowly over the water.

Sea Fog or **Advection Fog** is formed when warm, moist air moves over colder sea water. The moisture in the air condenses into fog, the same way a person's warm breath condenses on a cold window pane. The sea surface temperature must be cooler than the dew point temperature of the air. Unlike radiation fog, which requires calm or light wind conditions, sea fog may form when winds are moderate and may even persist as winds become strong. In fact, dense fog is often noted in Juan de Fuca Strait with westerly winds up to 30 knots. *Sea fog is most common in most areas during the summer and fall.*

Precipitation Fog forms in the vicinity of frontal systems when warm precipitation falls down through a cooler layer near the ground to saturate the surface air. Precipitation fog occurs mainly during winter storms.

Arctic Sea Smoke forms when very cold arctic air moves over warmer sea water. This is quite different from sea fog. In the case of arctic sea smoke moisture evaporates from the sea surface and saturates the air. The air is very cold, it cannot hold all of the moisture evaporated, so the excess condenses into fog. The result looks like steam or smoke rising from the sea surface and is seldom more than a few metres thick. Although sea smoke is normally not a hazard to mariners, under extreme conditions such as when strong winds blow down the inlets during an arctic outbreak, the fog may be thick enough and cold enough to create light vessel icing.

For most marine areas, the lowest chance of fog is in the spring. This however, is not the case for the inland waterways between Vancouver Island and the mainland, where the main type of fog is radiation fog. Typically radiation fog is quite rare in summer but becomes more frequent in the period from late autumn until early spring.

Snow

Snow also reduces visibility. It is usually a serious problem only in the mainland inlets, where it can occur mixed with rain. This condition results in greater radar attenuation than either rain or snow alone. As a result, not only is visibility seriously reduced, but also radar is less effective. This hazard is usually associated with arctic outbreaks.

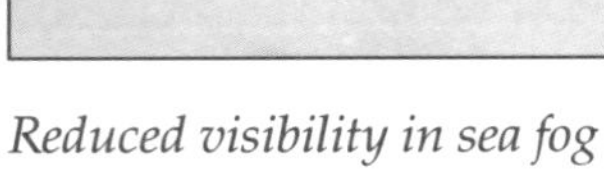

Reduced visibility in sea fog

Icing

Ice buildup on ships is a well-recognized hazard to both small and large vessels. Ice-coated decks threaten crew safety. Hatches, lifeboats and equipment become frozen and unuseable. Smaller vessels become top heavy and face the danger of capsizing when ice accumulation becomes severe.

Subzero temperatures are required for substantial ice buildup which only occurs on the B.C. coast during an arctic outbreak. This happens much more frequently over the North Coast marine areas than over the South Coast. At times, however, the arctic air pushes out to the west coast of Vancouver Island and conditions are right for icing. No area is free from this hazard and caution must be exercised during arctic outbreaks.

The eastern end of Dixon Entrance has a reputation for significant icing. Arctic air plunges out of Portland Canal and the strong winds and cold air temperatures provide ideal conditions for heavy vessel icing. Icing can result from freezing spray, freezing rain or fog. Its severity depends on the vessel, its speed, the wind speed, sea state and air temperature.

Ship icing can also occur in sea smoke which forms when very cold air flows over warmer water—this is a form of rime icing. The fog is composed of minute supercooled water droplets that freeze on contact with the ship. It is usually confined to a few metres above the water, but dense sea smoke conditions that resulted in measurable ice accumulations have been reported.

The M.V. ARCTIC PROWLER iced-up in winter.

Icing

Freezing Sea Spray is the most common form of ship icing. It occurs when the air temperature is less than –2° Celsius (the freezing temperature of seawater) and the wind is sufficient to produce blowing spray. The temperature of the sea water must generally be below 7° Celsius. Spray generated by the ship's motion also contributes to ice buildup. As the temperature drops and the winds increase the severity of the freezing spray will become greater.

During well established arctic outbreaks, strong outflow winds blow out most of the coastal inlets. These cold, strong winds are confined to relatively narrow jets as they leave the channel. A ship travelling along the coast could experience heavy ice accumulation on its windward side while crossing this region of strong winds. Once free of the icing zone, in the lee of coastal islands, the additional load of ice on only one side of the vessel could cause a heavy list.

The following nomogram shows the rate of icing accumulation for different wind speeds and air temperatures.

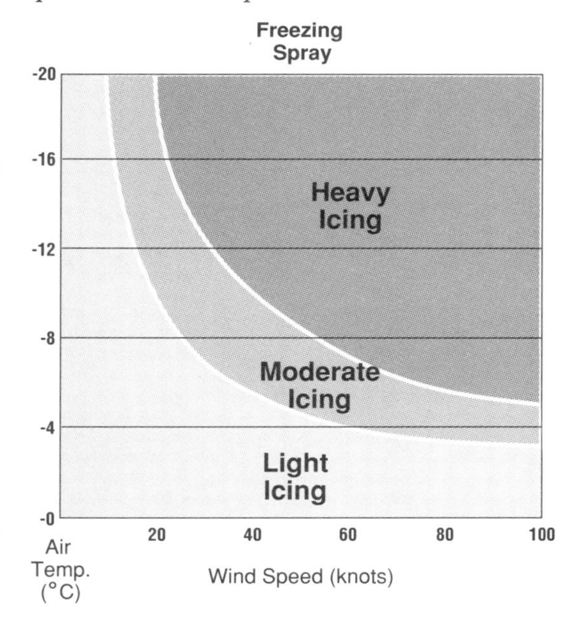

Icing conditions for vessels heading into or abeam of the wind

***Light icing**—less than .7 cm accumulation per hour*
***Moderate icing**—.7 cm per hour to 2 cm per hour*
***Heavy icing**—greater than 2 cm per hour*

Example:

When the winds are 50 knots and the air temperature is –10° Celcius, heavy icing can be expected.

"Heavy freezing spray" warnings are issued by the marine forecaster to alert the mariner of potentially dangerous icing.

Freezing Rain occurs when milder air arriving from the Pacific rides up over the cold arctic air on the coast. Rain from clouds in the warm air falls through the cold layer and becomes supercooled. Under these conditions the rain freezes when it hits the subzero temperatures of the vessel superstructure. A clear film of glaze is formed over the decks, railings and stairways. The amount of ice accumulated in freezing rain is usually not as great as it is with sea-spray icing.

CHAPTER SIX

Local Hazards

Local hazards are given for the following Marine Weather Forecast Regions

South Coast
- Strait of Georgia and Howe Sound
- Juan de Fuca Strait
- Johnstone Strait
- Queen Charlotte Strait
- West Coast Vancouver Island North
- West Coast Vancouver Island South

North Coast
- Queen Charlotte Sound
- Central Coast
- Hecate Strait and Douglas Channel
- Dixon Entrance East
- Dixon Entrance West
- West Coast Charlottes

There are no specific hazards listed for the two offshore areas, Bowie and Explorer. These areas frequently have high winds and waves associated with winter storms, but do not have topography or bathymetry influences to create localized hazards.

It is impossible to list every hazardous location along the British Columbia coast. But, by indentifying some specific problems the mariner should be able to identify other locations which have similar hazardous conditions.

A general description of the local weather is given for each of the coastal regions that Environment Canada issues marine forecasts and warnings (as listed on the preceding page). The order of the regions is the same as that found in the marine forecast. The locations of the local hazards are marked on the reference maps with an accompanying brief description of each hazard.

The symbols that are used to identify the local hazards are as follows:

Strong winds This includes all wind related hazards. The arrow shows the approximate wind direction.

High waves This includes all large waves which are formed by wind alone.

Steep waves These are waves which result from the interaction of two or more waves. This includes wave-current interaction.

Weather Hazardous weather such as fog and icing.

For most regions there is a graph, showing the frequency of occurrence of light to moderate winds (0–19 knots), strong winds (20–33 knots) and gale force and higher winds (34 knots or more). This graph is based on selected stations which are representative of wind conditions in the area and that have been in use long enough to provide a good data base.

General Conditions

The Strait of Georgia marine area extends from Orcas Island in the south to Quadra and Cortes Islands in the north. It includes the mainland fjords such as Jervis Inlet and the waters surrounding the Gulf Islands. Howe Sound is included in this chapter even though a separate forecast is issued for it.

Most of the weather hazards in the Strait of Georgia result from strong winds and waves interacting with tidal currents.

Wind circulation in the Strait follows the overall coastal pattern of northwesterly winds in summer and southeasterly winds in winter. During winter storms, strong southeasterlies develop in advance of a front. In most cases winds slacken when the front passes, but when a ridge of high pressure builds behind the front, strong northwest winds develop in the Strait. The northwest-southeast orientation of the Strait and Coast Range mountains channel the wind between the mainland and Vancouver Island. Some of the strongest winds recorded are northwesterly gales over the southern part of the Strait.

In winter, arctic air from the interior surges through mountain passes and down the fjords of the coastal mountains giving strong outflow winds. The northerlies from Howe Sound called "Squamishes" can extend across the Strait of Georgia to the Gulf Islands.

Outflow winds from the Fraser Valley can similarly affect the extreme southern part of the Strait of Georgia. Gale force winds can extend across to Saturna Island, however, this is rare. In general, outflow conditions produce the strongest winds through the inlets. Gale force winds in the inlets generally ease to about 25 knots over the open water outside the inlets.

In the summer weak frontal systems pass over the area and often give little or no cloud. If pressures rise strongly behind the front, strong northwest winds develop which may prevail for two to three days.

The sea breeze is an important consideration in summer. On the east side of the Strait of Georgia the sea breeze augments the prevailing westerly winds and speeds reach 20 knots or more during the late afternoon. At night drainage winds are much lighter. Caution is required in some coves where winds funnel down from mountain valleys producing sudden gusty conditions in the evening. During the summer months, most of the mainland inlets will regularly experience southerly winds during the day and northerly winds at night.

With southeast winds the highest waves develop north of Entrance Island and can build in extreme cases to nearly 4 metres. Near Texada Island, tides and winds are generally stronger due to channelling between land masses. The water in this area is also quite shallow and as a result the southeast wind waves steepen on the currents making sailing conditions quite hazardous for small craft. Under northwest winds, the highest waves will be found in the southern part of the Strait.

The Strait of Georgia is an inland sea and, except over the extreme southern part where it joins Juan de Fuca Strait, it is not influenced by sea fog. The best visibilities occur in the summer months. During winter, visibilities are lowered in rain and fog that accompany winter storms. In fall and sometimes in summer, westerly winds bring sea fog into Juan de Fuca Strait which occasionally reaches the southern part of the Strait of Georgia.

A large mass of fresh water enters the Strait of Georgia from the Fraser River. The river water moves across the Strait on top of the heavier salt water. When winds exceed 15 knots this surface layer will tend to move downwind. For speeds greater than 25 knots, the surface currents will be driven mainly by the wind. Tides will have a secondary effect. Wind speeds should be considered when estimating tidal currents in the vicinity of the Fraser River estuary.

The difference in density between salt and fresh water can cause buoyancy problems for ships going from the Strait of Georgia into the Fraser River. A ship which is loaded to the gunwhales when in the Strait of Georgia will suddenly find the decks awash as its hull rides deeper in the fresh water of the river.

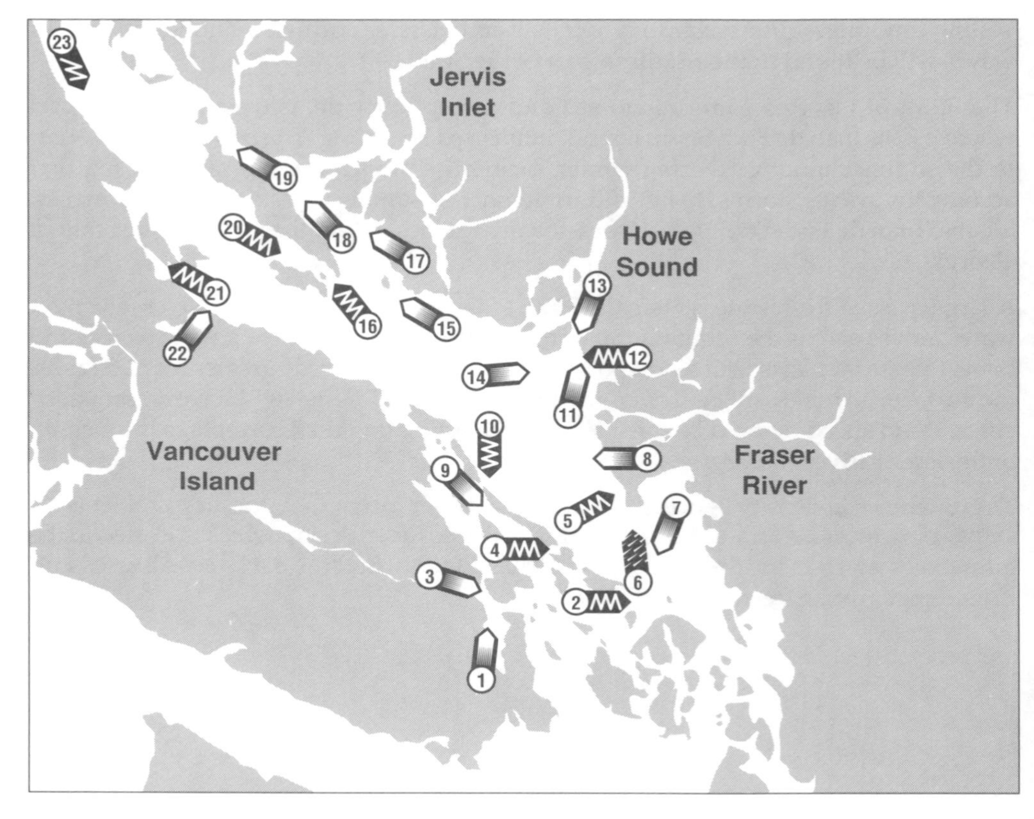

1. Strong winds Strong southerly winds can blow through Squally Reach and the Saanich Inlet. These winds occur after a fast-moving cold front moves over southern Vancouver Island and a ridge of high pressure builds rapidly behind it. While these winds do not happen often, speeds up to 40 knots can occur.

2. Steep waves Steep, confused seas in Boundary Pass result from easterly winds blowing against the flood tide currents. These conditions can be severe near East Point and quite dangerous to small craft. Tide rips add to the hazard. East Point Lighthouse reports are representative of strong easterly winds over the southern Strait of Georgia.

3. Strong winds Strong westerly winds blowing out of Cowichan Valley and through Satellite Channel can be quite dangerous to small vessels. These are "Qualicum" type winds. Gusty night time drainage winds also occur.

4. Steep waves Large flood tides through Active Pass form rips near the eastern entrance. Northwesterly winds and waves create steep, rough seas in a narrow band just outside the entrance to the Pass as they counter the current.

5. Steep waves Southwest winds and waves moving onto Roberts Bank and against ebb currents create steep rough seas. Also, flood tides tend to flow onto the bank. Surface currents can be increased by onshore winds. These conditions make navigation over the shallow waters very hazardous during bad weather. Sand Heads' wind reports are representative for this area.

6. Fog Westerly winds over Juan de Fuca Strait can carry sea fog into the southern part of the area, often quite suddenly. After a spell of hot weather, a change to cool sea air is usually preceded by an increase in winds, to south 15 to 20 knots, during the morning at Sand Heads. This change often provides warning of foggy conditions.

7. Strong winds Northeast outflow winds, exceeding 40 knots, from the Fraser Valley produce hazardous conditions across the southern Strait of Georgia. The East Point report can indicate overwater-wind strength. Air temperatures rarely fall low enough for icing but the potential exists. In early February 1989 severe icing problems occurred.

8. Strong winds The surface layer of the Fraser River plume is affected by strong winds, resulting in unexpected currents. The plume may also contain logs, trees and other flotsam that can be hazardous to small vessels.

9. Strong winds Northwest winds funnel into Trincomali Channel producing winds much stronger than in the Strait of Georgia.

10. Steep waves Flood-tide currents set from Trincomali Channel through Porlier Pass into the Strait of Georgia at about 9 knots maximum speed. Northwest winds against this current creates steep, rough seas just outside Porlier Pass. Waves generated by easterly outflow winds from the Fraser Valley in winter, steepen on the flood currents, creating hazardous conditions for small vessels.

11. Strong winds Tidal streams from Burrard Inlet and Howe Sound converge at Point Atkinson producing tide rips. Passage can be hazardous to small vessels during fresh to strong, southerly winds and short waves. Point Atkinson Lighthouse reports are representative for south to southwest winds.

12. Steep waves A short chop may develop from Point Atkinson to First Narrows when westerly winds oppose strong ebb tide currents.

13. Strong winds Outflow winds in Howe Sound with speeds up to 40 knots—**the "Squamish"**—occur in winter. Wind speeds generally increase from Squamish to the mouth of the inlet and then spread out in a jet over the Strait of Georgia. The Pam Rocks buoy is a good indicator of Squamish winds, but Point Atkinson is not, due to sheltering. See Squamish wind diagram on p. 30.

14. Strong winds On clear, sunny days in summer westerly sea breezes can reach 20 to 25 knots by late afternoon. Winds die down in early evening and remain light over night.

15. Strong winds Southeast winds in the Strait of Georgia are often strongest along the Sunshine Coast. Representative wind speeds will be reported by Merry Island Lightstation. These winds are strengthened in Welcome Passage due to funnelling between cliffs on either side of the channel. Southeast waves will steepen on ebb currents through Welcome Passage creating rough seas.

16. Steep waves Short, very steep seas form to the southeast of Lasqueti Island under southeast winds.

17. Strong winds Southeast gales cause gusty winds in the lee of coastal mountains affecting Pender Harbour. Merry Island Lightstation reports are indicative of gale force southeasterlies in this area.

18. Strong winds Strong southeast winds in Malaspina Strait concentrate at Cape Cockburn producing rough seas.

19. Strong winds Southeast wind is accelerated around Grief Point giving speeds 5 to 10 knots higher than the area forecast.

20. Steep waves Rough seas occur in Sabine Channel when strong southeasterly or northwesterly winds are blowing against tidal currents. Combined wind, wave, and current conditions can be hazardous for small craft.

21. Steep waves Tidal currents stream at 2 to 3 knots at the southern entrance to Baynes Sound. Waves steepen to give rough seas on strong southeasterly winds against the ebb tide setting south.

22. Strong winds Strong southwesterly winds blow over Vancouver Island descending onto the Strait of Georgia—these are known locally as "Qualicums." A narrow jet of strong wind extends from Qualicum Beach to the northern tip of Lasqueti Island and enters False Bay. These winds build to a maximum of 30 to 40 knots during summer afternoons. Reports of southwest winds from the Sisters Island Lightstation indicate when "Qualicums" are blowing.

23. Strong winds and high waves Tidal currents from the southern Strait of Georgia and from Discovery Passage converge near Sentry Shoal, just south of Mitlenatch Island. Strong, southeast winds generate waves that steepen on the tide rips. The rough seas, strong currents and winds can be hazardous to small craft.

Winds – Sisters Island

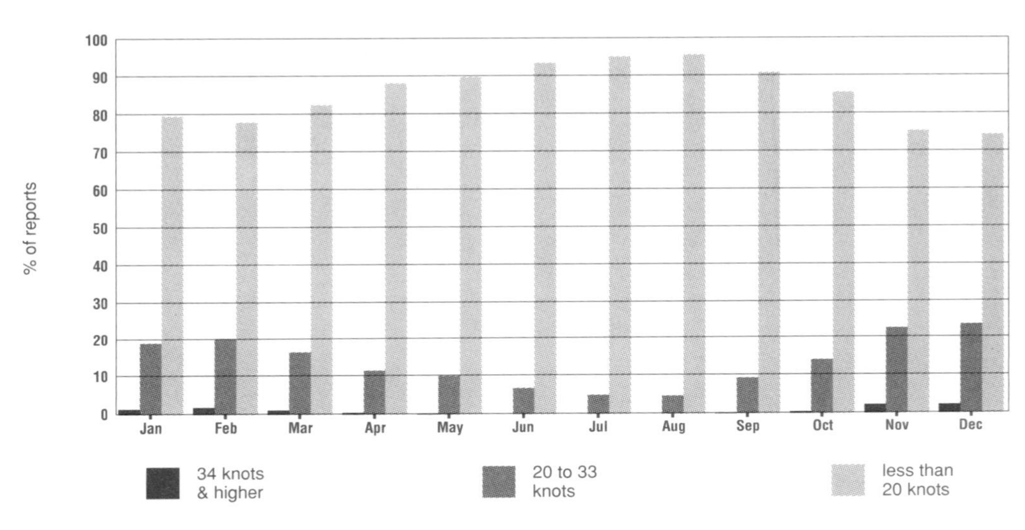

Winds – Sand Heads

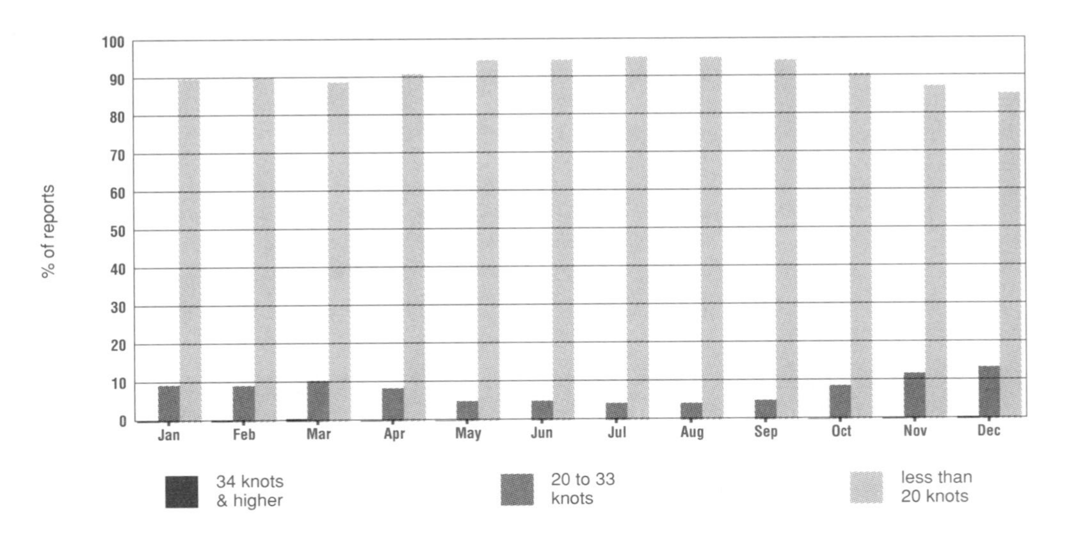

General Conditions

Fog presents one of the most severe hazards over the western part of Juan de Fuca Strait. Sea fog forms frequently in summer and fall and is carried over the Strait by the prevailing westerly winds. Poor visibility, lowered to half a mile or less, can be accompanied by winds of 25 knots.

Strong winds, often accompanied by fog and drizzle, present the other major hazard to mariners. The winds through Juan de Fuca Strait are generally not that strong just ahead of an approaching Pacific storm even through the winds further offshore reach gale force. The winds at the entrances to the Strait, however, often rise to gale force southeasterlies. Strong southeast winds frequently blow out of Puget Sound, across the eastern entrance of the Strait and up into Haro Strait during a winter storm. Discovery and Trial Islands will report gale and sometimes storm force winds in this situation. The most intense storms will occur when a coastal low moves from the southwest directly over southern Vancouver Island.

Following the passage of a front however, the winds shift to a westerly direction and frequently penetrate the Strait at gale-force strength. For major storm systems, the strongest winds in the Strait will usually occur a few hours after the front has passed and not ahead of it. Elsewhere over the coast the maximum winds generally occur ahead of the front.

In summer, the sea breeze is an important factor in the wind regime. Under normal conditions an area of low pressure lies over the B.C. interior and winds in the Strait are westerly. The winds strengthen during the day to 25 or 30 knots by late afternoon but gradually subside during the late evening. When pressures are particularly low in the interior, wind speeds may reach 40 knots.

On some occasions in summer and fall, the trough of low pressure from the interior moves westward to lie just off the coast. Winds in the Strait turn to light easterlies and skies are sunny with above normal temperatures. This weather pattern may continue for several days until the low pressure trough suddenly moves back to its usual position in the interior. Winds then shift to westerly 20 to 30 knots accompanied by fog and drizzle.

Lighthouse reports are useful for monitoring weather in the area. Winds at Race Rocks are representative of westerly winds. Both Trial Island and Discovery Island lightstations give a good indication of strong easterlies over the eastern portion of the Strait.

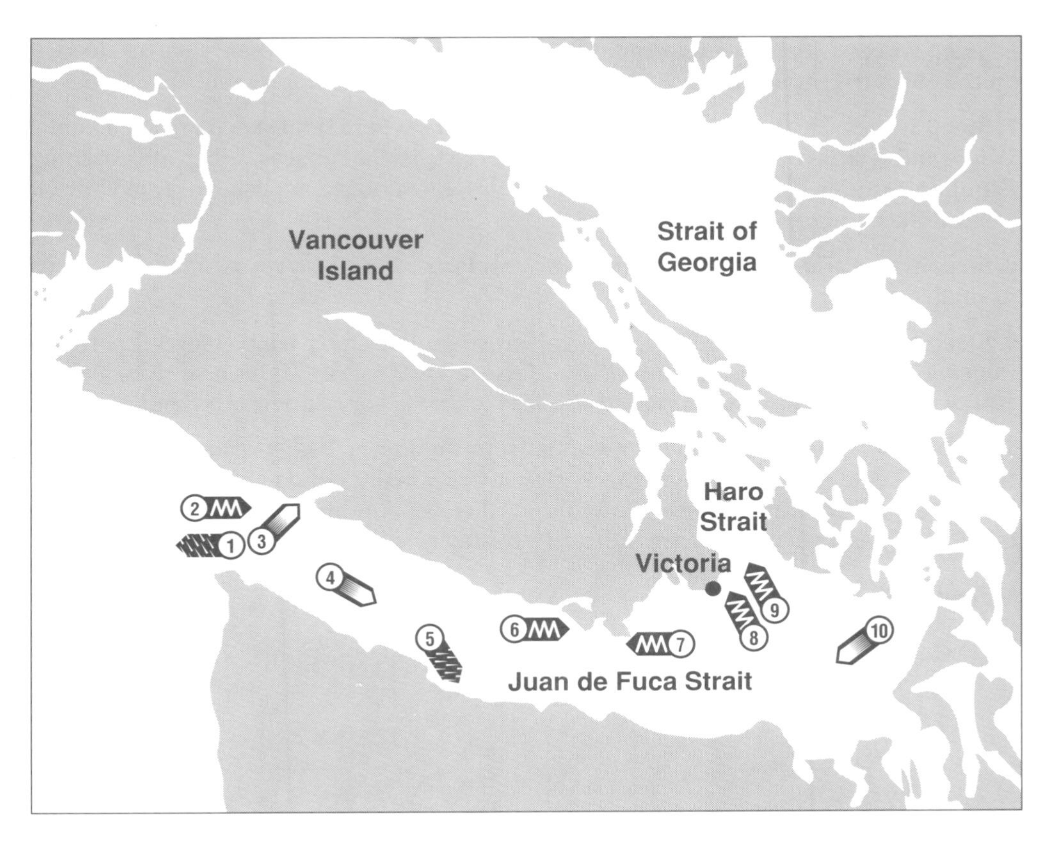

1. Fog Sea fog is prevalent in the western entrance during late summer and fall. Combined with heavy seas and strong winds, sea fog poses a hazard to all mariners.

2. Steep waves Swell is prevalent in the western entrance. On ebb tides the waves steepen forming short choppy seas that are potentially dangerous to small craft.

3. Strong winds Port San Juan is exposed to southwesterly gales and heavy seas. Caution is warranted when entering the inlet under these conditions.

4. Strong winds In the summer, after a period of light easterly winds and warm temperatures, westerly winds rising to 35 to 40 knots can abruptly surge through the Strait. There are few visual indications to warn of this change. Listen to the marine forecast and warnings.

5. Fog Fog is frequently found over the southern shore of Juan de Fuca Strait at many times of the year.

6. Steep waves Tidal streams around Beechey Head form rips. Westerly waves steepen on the ebb currents (exceeding 3 knots) producing a short, choppy sea.

7. Steep waves Heavy tide rips occur at Race Rocks where tidal speeds reach 6 knots. Waves and winds countering the current produce steep chaotic seas. When approaching Esquimalt under stormy conditions, small vessels should exercise extreme caution around Race Rocks.

8. Steep waves Tidal streams around Discovery Island form heavy rips, often dangerous to small craft. Easterly winds and choppy waves add to the hazard.

9. Steep waves During Winter storms, gale force southeasterly winds blowing through Puget Sound and into Haro Strait can build seas up to 3 metres. Tides near the southern end of Haro Strait can also oppose these seas to create very dangerous conditions.

10. Strong winds Outflow winter winds from the Fraser Valley bring northeasterly winds and cold temperatures to this region. Occasionally winds will reach gale force, with lowered visibility in snow on windward coasts. Lightstation reports from Trial Island and Discovery Island are indicative of strong winds and weather in the area.

Winds – Race Rocks

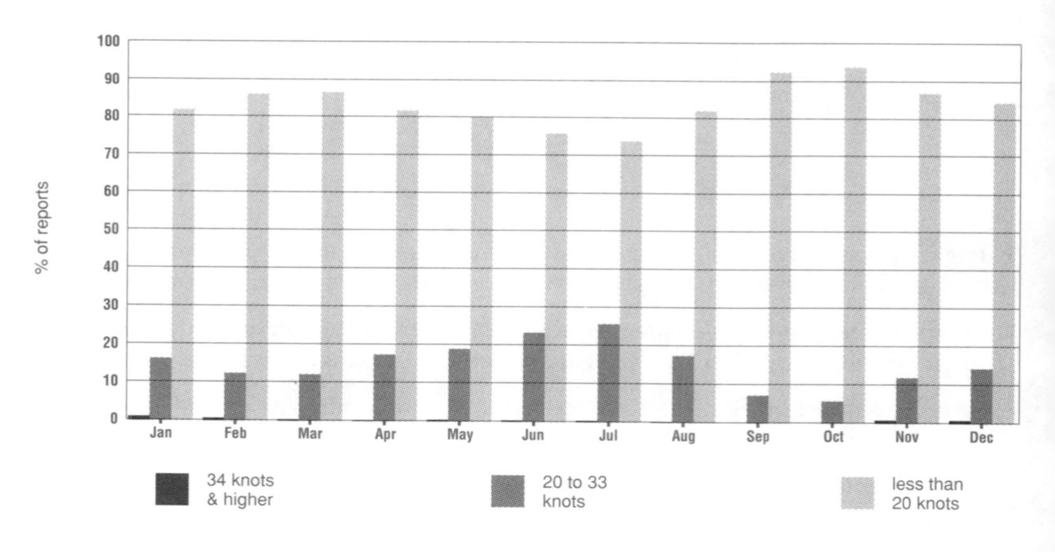

General Conditions

The Johnstone Strait marine area includes Johnstone Strait and Discovery Passage. The many inland waterways from Bute Inlet to Desolation Sound are also included in this region, but due to the complexity of all the islands and channels it is impossible to give a simple forecast statement of the overall wind conditions in this area. The mariner who frequents these inland waterways must take the winds mentioned in the forecast for the Strait of Georgia and Johnstone Strait and modify them, through local knowledge, for the inland waters.

Southeasterly winds are the strongest during the winter, with the westerlies more predominant in the summer. Johnstone Strait experiences more storm force winds during the winter than in Discovery Passage. The winds are strongly funnelled between the bordering mountains and can be as much as 15 knots stronger, in some situations, than the winds reported at observing stations in Queen Charlotte Strait.

Southeasterly winds do not usually create high seas through the Strait from Chatham Point to Yorke Island, choppy seas result. The most hazardous condition occurs when westerly winds are blowing against an ebb tide.

In general, conditions can be very dangerous when the wind opposes the tides.

In the summer, westerly winds develop in Queen Charlotte Strait during the afternoon as a light sea breeze. It is thought that the winds are funnelled down Johnstone Strait, moving like a wave, and can reach 30 to 35 knots at Chatham Point by evening.The winds usually ease by around two in the morning. Chatham Point is often the only reporting station that will record these winds.

Near Cape Mudge, at the southern entrance to Discovery Passage, the conditions are particularly hazardous and warrant extreme caution when strong southeast winds blow up the Strait of Georgia. These winds generate short waves that steepen on the south flowing tidal currents to produce very rough seas. The combined effect of waves, wind and current is dangerous for small vessels.

The Helmcken Island wind report is fairly representative for southeasterly winds but does not record the full strength of westerly winds. The Chatham Point lightstation records the westerly winds through the Johnstone Strait though not always their full strength and is not representative of the winds in Discovery Passage.

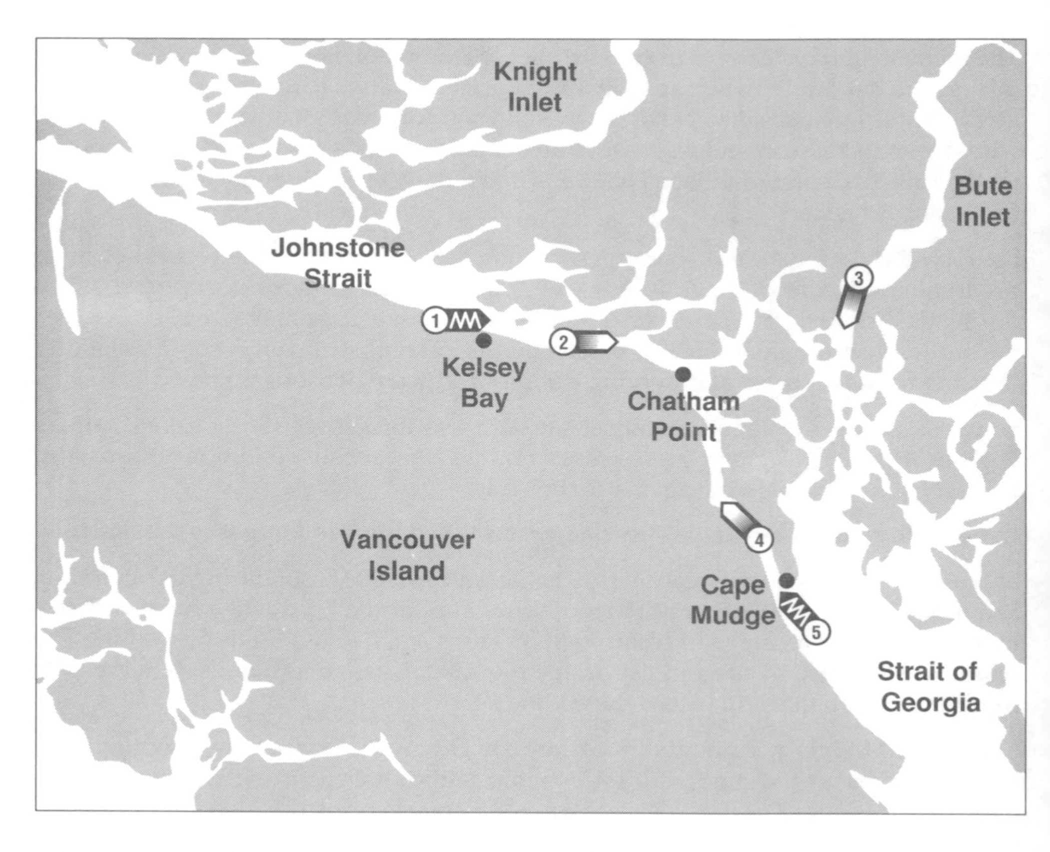

1. Steep waves A westerly wind opposing the ebb tide can create dangerous tidal rips in the passages south of Hardwicke Island, near Kelsey Bay. When winds of 30 to 40 knots are blowing in the same direction as the tides, the seas in this area may only build to around one metre. However, when these same winds are blowing against the tides the seas can build to 3 to 4 metres.

The best advice is to travel when the winds are in the same direction as the tides. With westerly winds, travel with the flood tide.

2. Strong winds Sea breeze winds which originate in Queen Charlotte Strait intensify as they funnel through Johnstone Strait. They can reach 30 to 35 knots during summer evenings. These winds continue, although with reduced strength, through Discovery Passage, and also into Hoskyn and Sutil Channels. The winds diminish overnight.

3. Strong winds Gale-force outflow winds can be expected in Bute Inlet during arctic outbreaks in winter. Freezing spray is possible if air temperatures fall a few degrees below zero. The full force of arctic outflow winds is not often felt in Johnstone Strait itself but can be very strong in some of the other channels which lead down from the mainland mountains.

4. Strong winds Strong, southeast winds are intensified in Discovery Passage. This can lead to hazardous conditions for small craft on opposing tidal currents. Tidal currents range up to 10 knots in the channel.

5. Steep waves Tidal streams off Cape Mudge reach 5 to 7 knots, flooding south and ebbing north. Between Cape Mudge and Willow Point there is a heavy race on flooding streams. Opposing southeast winds set up short steep waves which steepen further on the current creating rough, breaking seas. The sea can be very dangerous to small craft. Passage at and immediately after high water slack is recommended.

In the Johnstone Strait marine forecast area, weather information is available from the following sources:

An aerial photograph of the coast looking up Bute Inlet.

General Conditions

Winds in Queen Charlotte Strait are predominantly from the southeast in winter. The strongest of these winds is associated with the frontal passage in storms. Winds tend to be from the northwest in summer.

Hazardous wave conditions are found in heavy northwest swell which can penetrate into the western section of Queen Charlotte Strait under northwesterly winds. These winds occur following storms in winter as a high pressure ridge builds behind the front.

During cold arctic outbreaks, gale force winds and low temperatures can produce freezing sea spray in most of the major mainland inlets. Under very strong winds with cold temperatures the smallest sea spray particles freeze in the air forming ice crystals. This ice-crystal fog reduces visibility to near zero.

Often outflow winds extend from the mainland inlets over Queen Charlotte Strait. In traversing the area, vessels will experience abrupt changes in wind speed. Gale force winds at inlet openings may fall off greatly in sheltered areas between fjords and channels.

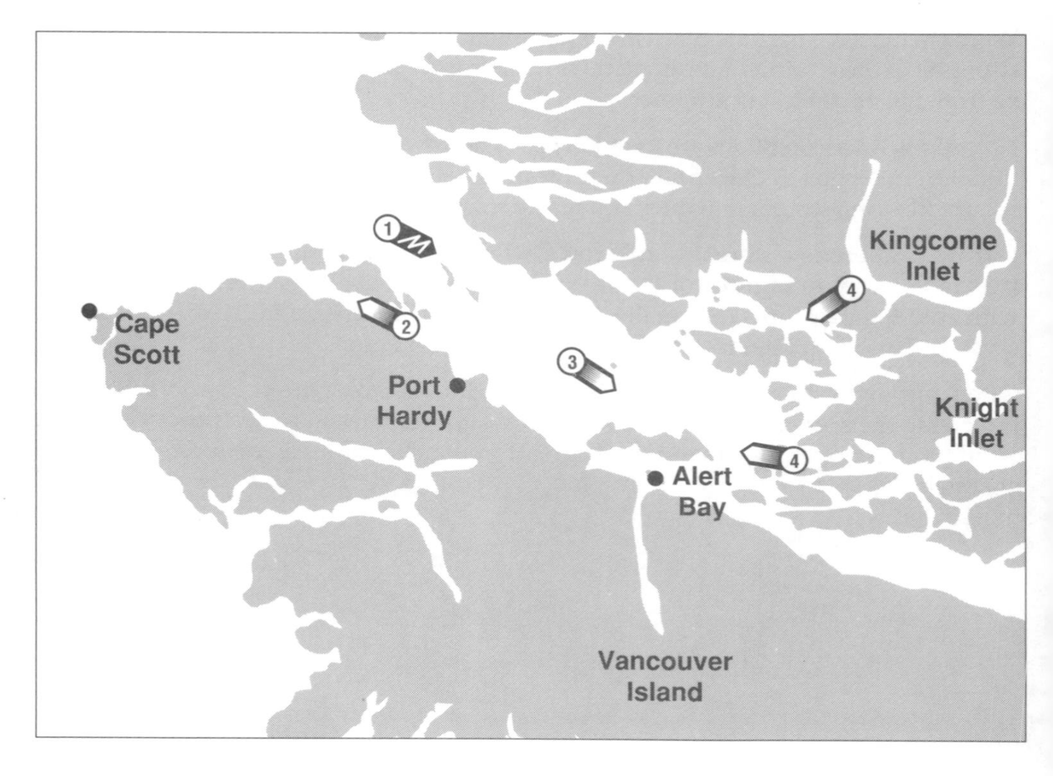

1. High waves The western entrance to Queen Charlotte Strait is open to the Pacific Ocean. Heavy swell under westerly winds can frequently make conditions hazardous to small craft.

2. Strong winds Strong southeast winds are funnelled in Goletas Channel increasing speeds in this area.

3. Strong winds On clear sunny days during summer months a sea breeze generally develops in late morning building to 25 to 30 knots by late afternoon. The sea breeze dies out by evening to be replaced by a weaker land breeze at night. Winds are strongest on the east side of the Strait.

4. Strong winds Outflow winds from Knight Inlet and Kingcome Inlet during arctic outbreaks blow down across Queen Charlotte Strait. These can lead to gale force winds over limited areas of the strait which are dangerous due to their strength and suddenness. Sea-spray icing is possible in air temperatures below –5° Celsius. Reports from Port Hardy and Alert Bay of easterly winds, indicate outflow wind conditions but the strength of these winds are usually much lower than the wind conditions in the main channels of the inlets.

Winds – Herbert Island

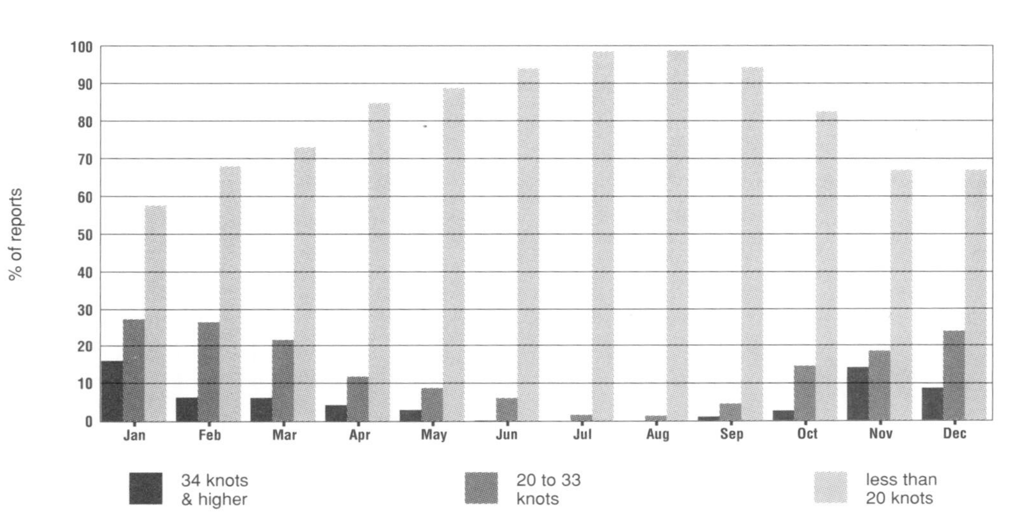

General Conditions

Along the outer coast the winds are predominantly from the southeast in the winter and from the northwest during the summer.

The strongest winds accompany intense lows and their associated fronts. Generally, as a low approaches the coast, winds back into the southeast and increase in speed, often reaching gale or storm force. With the passage of the front, winds usually shift to the northwest. Sometimes periods of strong southwest winds occur, but they are usually much lighter than the southeast or northwest winds. In the most intense winter storms, the maximum sustained wind speeds can rise to 70-80 knots with gusts up to 95 knots.

The steepness of the terrain influences both the speed and direction of the winds to a significant degree. In any given locality speed and direction may depart widely from the general pattern. This is particularly so in inlets not aligned with the general wind direction, such as Quatsino Sound.

In winter, winds of gale force or stronger, occur quite frequently. The duration of individual storms is usually about 1 day while the interval between storms varies from about 1 to 5 days. However, under exceptionally settled winter conditions, periods of up to 2 weeks may elapse without gales. In summer, storms are infrequent and less intense than those of winter. The interval between summer gales can vary between 2 to 6 weeks. Autumn and spring are transitional seasons between the more settled weather of summer and the stormy winter weather. Equinox storms often mark the beginning or end of the stormy season.

Strong northwest winds which rise at times to gale force 35 to 40 knots are fairly common off the coast of northern Vancouver Island in summer. These winds occur when a lee trough becomes established over the southern section of the Island. Solander Island and occasionally Sartine Island will record these winds.

When arctic air moves southward over the province in winter, it may cross the coastal mountains and continue westward to invade the west coast of Vancouver Island. Strong outflow winds may develop in channels and inlets that extend back into the mountains. Outflow winds often blow against an opposing westerly swell creating chaotic seas. Sea spray icing will occur when air temperatures fall to a few degrees below zero. Snow showers offshore can lower visibility.

The passage around the north end of Vancouver Island is one of the most dangerous in B.C. waters. Tidal currents range in speed up to 3 knots or more in Scott Channel. East of the Scott Islands currents are slightly weaker but produce rips and surface overfalls. Heavy seas and strong winds, mainly from the southwest and west, combine with currents to create extremely dangerous conditions for small vessels.

Sailing Directions recommends that this part of the coast should be given a wide berth.

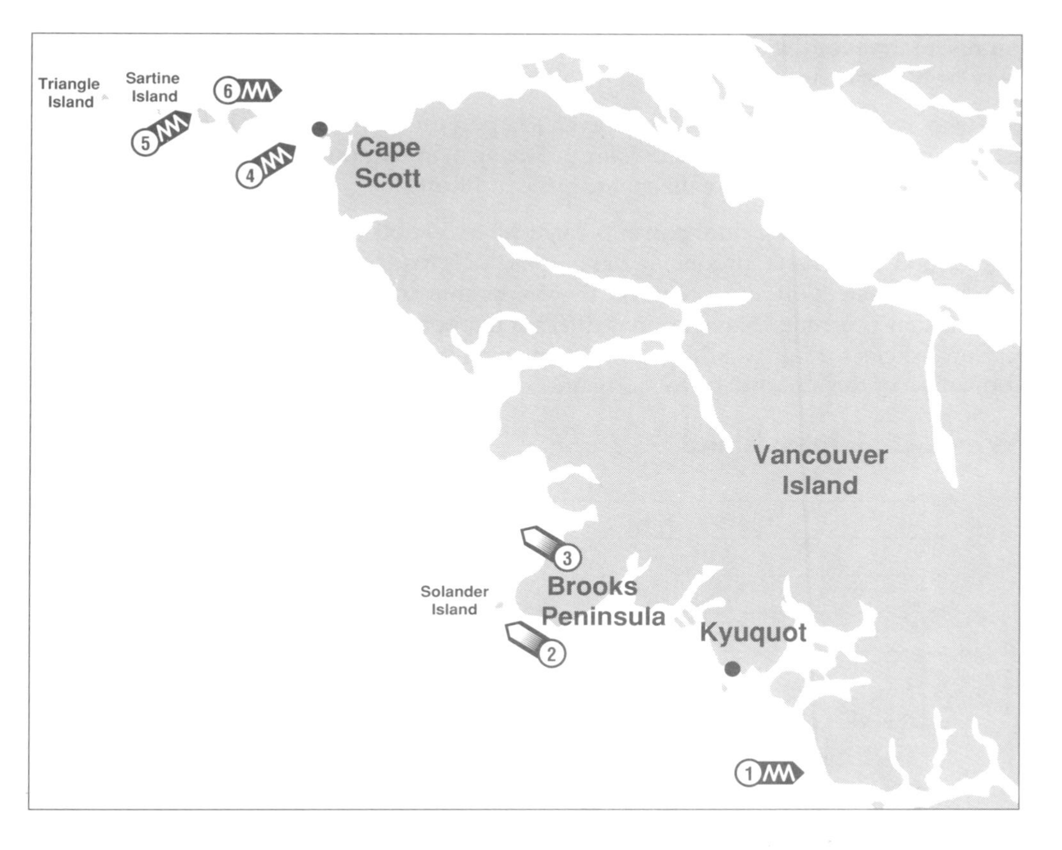

1. Steep waves Westerly waves steepen on tidal currents around Tatchu Point. Conditions are hazardous under strong winds and heavy seas.

2. Strong winds and steep waves The passage around Brooks Peninsula under storm conditions is one of the most hazardous on the west coast. Winds parallel to shore accelerate around the peninsula due to the blockage created by the mountains. Winds off Solander Island may be 15 knots higher than further offshore or away from the point. Solander Island reports are representative of wind speeds near the headland. Wind waves and swell steepen on tidal currents.

3. Strong winds Lee winds behind Brooks Peninsula are very gusty during southeast gales. These conditions can be hazardous to smaller vessels seeking shelter in Brooks Bay.

4. Steep waves Ebb-tide currents of 3 knots in Scott Channel, which set to the southwest, interact with the southwesterly wind waves and swell. This results in waves which steepen and break and can be dangerous to small craft. Cape Scott wind reports are representative for the approaches to Scott Channel except for southeast winds, which may be 10 knots stronger over water.

5. Steep waves Southwest wind waves steepen on tidal currents between Triangle Island, Sartine Island and Lanz Island. Strong winds, poor visibility and confused seas over the tidal flows create hazardous conditions.

6. Steep waves Strong tidal currents from 1.5 to 3 knots are present around the Scott Islands and Cook Bank, producing dangerous tidal rips and overfalls. Westerly swell and wind waves steepen and break on these currents compounding the danger of the overfalls and currents. Navigation is difficult in poor visibility. Mariners are advised to plan passages for fair weather using the marine forecasts. Cape Scott reports give an indication of the strength of westerly and northwesterly winds over Cook Bank.

Winds – Solander Island

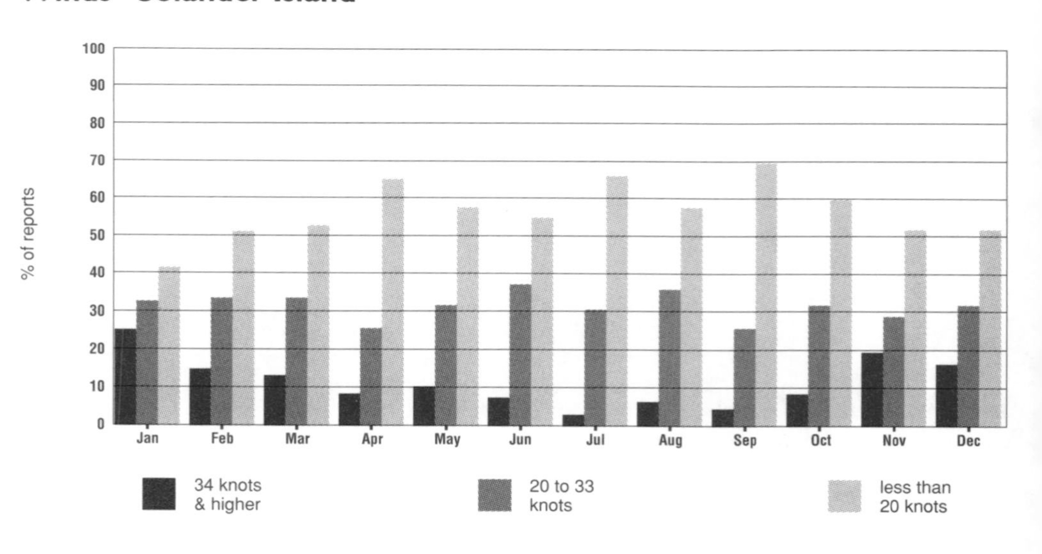

General Conditions

The southern section of West Coast Vancouver Island is influenced, as is the northern section, primarily by the approach of intense lows and frontal systems. The winds are mainly from the southeast or northwest as they are forced to follow the main orientation of the Vancouver Island mountains.

Frontal systems normally move first onto the north coast and then gradually weaken as they move across Vancouver Island. By the time the front reaches southern Vancouver Island the associated winds are much lighter than they were in the north. Southeast gale force winds often occur during the winter, but storm force winds (48-63 knots) only occur during the strongest of winter storms. The strongest southeast winds generally occur when a low approaches Vancouver Island from the southwest, with the front moving directly eastward over the island.

Northwest winds develop when a ridge of high pressure builds rapidly towards Vancouver Island after the passage of a front. In this situation northwest winds, often up to gale force, will spread down the coast and into the entrance of Juan de Fuca Strait.

If the ridge builds further eastward over the mainland coast, the winds over the northern end of the island veer toward the northeast. As this happens a lee trough develops over the southern section of Vancouver Island. With a well developed lee trough, the winds close to the coast diminish and become light and variable. The area of light winds often extends right along the coast up to the Brooks Peninsula. At other times, when the lee trough is not as well established, the area of light winds will be only found south of Estevan Point. Strong northwesterly winds can still occur further offshore, outside the zone of these light lee trough winds.

When there is a broad trough of low pressure to the west of Vancouver Island, with light winds near the coast, an event called a stratus surge can take place. This event is described in the section on STORMS, p 37. During a stratus surge light easterly winds can increase very quickly to strong and occasionally gale force southerlies. Visibility can drop suddenly with the arrival of stratus and fog.

In winter, during an arctic outbreak, strong north to northeast winds blow out through the many inlets leading to the coast.

Mariners are advised that tidal currents tend to set towards the coast. The "current set" is accentuated by the strong inflow into large sounds and by strong south-southwesterly winds. Tidal streams are often very strong near headlands and in inlets. Storm generated waves and swell steepen on these currents, adding to the danger.

Northwest winds in summer produce upwelling of cold water along the coast. Dense sea fog occurs as the warmer air passes over the cold water. Visibility is worst in August occasionally falling below 2 miles. Visibility is best in the spring.

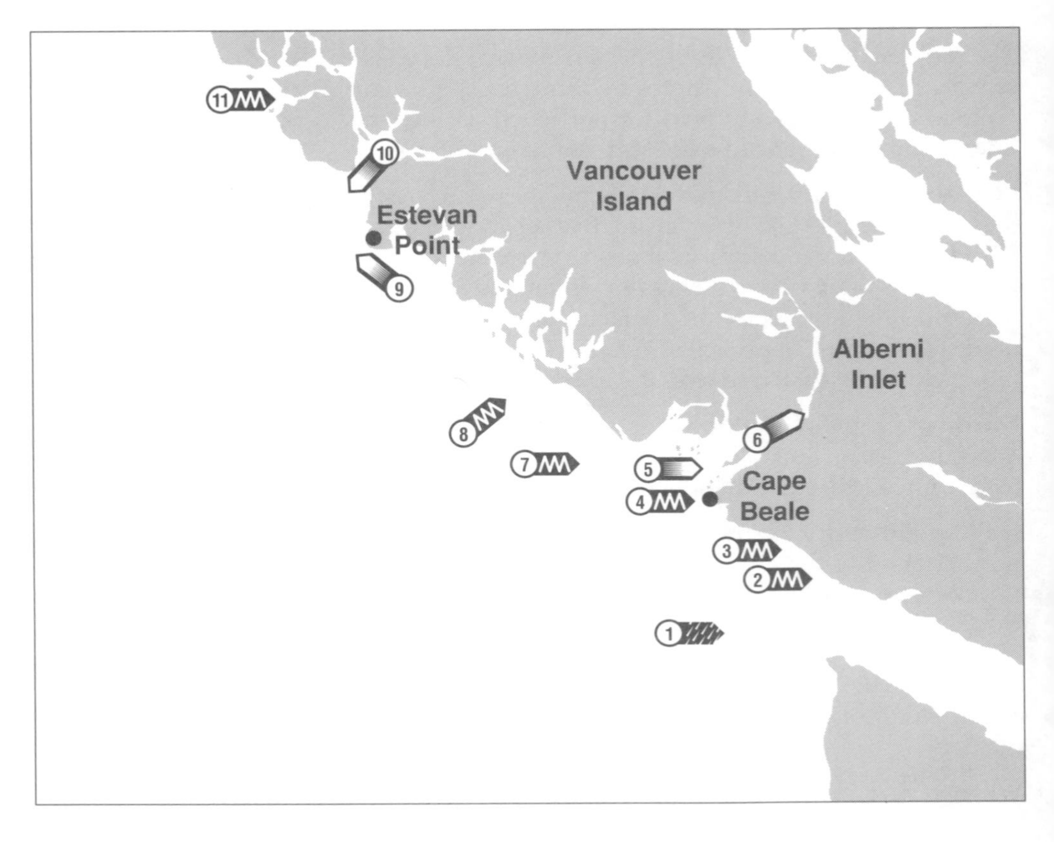

1. Fog Sea fog is prevalent over the entire area but is worst in the western entrance to Juan de Fuca Strait.

2. Steep waves Westerly seas steepen and break on ebb tidal currents about one mile off Bonilla Point. Combined northwesterly winds and steep waves are hazardous to small vessels. In foggy conditions this area is particularly hazardous.

3. Steep waves Westerly swell, tides and the river flow from Nitinat Lake combine to produce steep choppy waves. The rocks and sandbars of the Nitinat River bar add to the hazards.

4. Steep waves Strong tidal currents and rips around Cape Beale steepen waves to breaking. Rounding Cape Beale into Trevor Channel is hazardous whenever strong winds and heavy seas counter the ebbing currents. Cape Beale Lighthouse reports are representative of overwater conditions and can also indicate outflow winds from the inlet.

5. **Strong winds and steep waves** Westerly winds exceeding 25 knots and waves steepening on ebb tide currents in Barkley Sound create shock-like, confused seas.

6. **Strong winds** Southwest winds are funnelled through Barkley Sound and up Alberni Inlet causing strong winds in the inlet. Winds blowing in the direction of the tidal flows in Alberni Inlet can increase the surface currents.

7. **Steep waves** Refracted swell and heavy wind waves over La Perouse Bank cross near the mouth of Barkley Sound producing confused seas. Waves may steepen and break on ebb tide currents adding to the hazard.

8. **Steep waves** Southwest swell crosses with southeast-wind waves during strong winds and gales ahead of storm fronts. Confused seas result.

9. **Strong winds and steep waves** Southeast winds offshore which can reach 60 knots in storms are generally indicated by the gusts reported by Estevan Point. Rounding Estevan Point in strong winds and heavy seas or swell is hazardous as the south-southeast waves steepen on the strong tidal currents close to the point.

10. **Strong winds** Gale-force outflow winds occur in Nootka Sound in winter. Locally generated waves interact with west-southwest swells to produce chaotic seas. Snow can lower visibility and sea spray icing may occur when air temperatures fall below minus 5° Celsius. Nootka Lighthouse reports are representative for winds and visibility during cold air outflows.

11. **Steep waves** Westerly waves steepen on strong ebb tide currents in Nuchatlitz Inlet creating confused seas.

Winds – Estevan Point

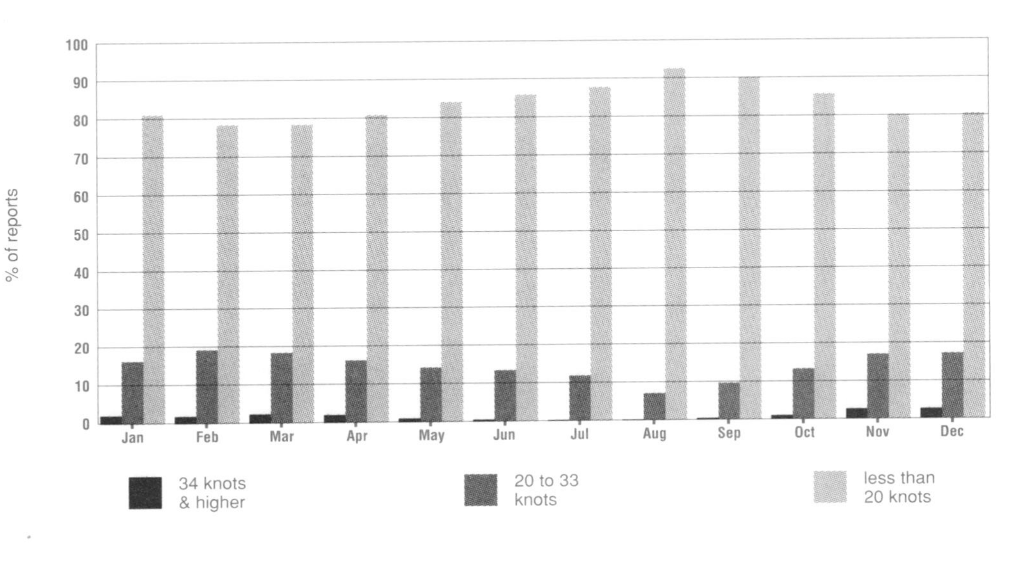

General Conditions

Winter storms which produce extreme winds and high waves are the principal hazards in Queen Charlotte Sound. Coastal lows crossing the Charlottes produce the worst conditions. Storm force winds and seas of 8 to 10 metres can be expected several times each winter over the Sound.

Close to the coastline, seas interact with strong tidal currents. As the waves steepen and break, they may swamp or broach smaller vessels. The area off Cape St. James and the Scott Islands are notorious in this respect.

Swells out of the south-west quadrant, with periods from 16 to 22 seconds are refracted over the banks and shoals in Queen Charlotte Sound resulting in chaotic, rough and heavy seas. Because swell directions and periods can change over a few hours, unexpected rough seas may be encountered over short distances but they will be transitory and will move away quite quickly. In autumn and winter, mariners can expect swell exceeding 4 metres (mainly from the west or southwest) about 50% of the time.

The drilling rig SEDCO 135F, working off Cape St James, reported an extreme wave nearly 30 metres high on October 22, 1968. This is the largest wave ever recorded on the coast. It was not only the extreme magnitude of the wave which was dangerous but also the speed in which the seas developed. At 7 am the combined sea height was only 3 metres—it rose to 18 metres in just 8 hours.

Visibility in Queen Charlotte Sound is best in the spring. In winter, visibility is reduced by precipitation and fog accompanying winter storms.

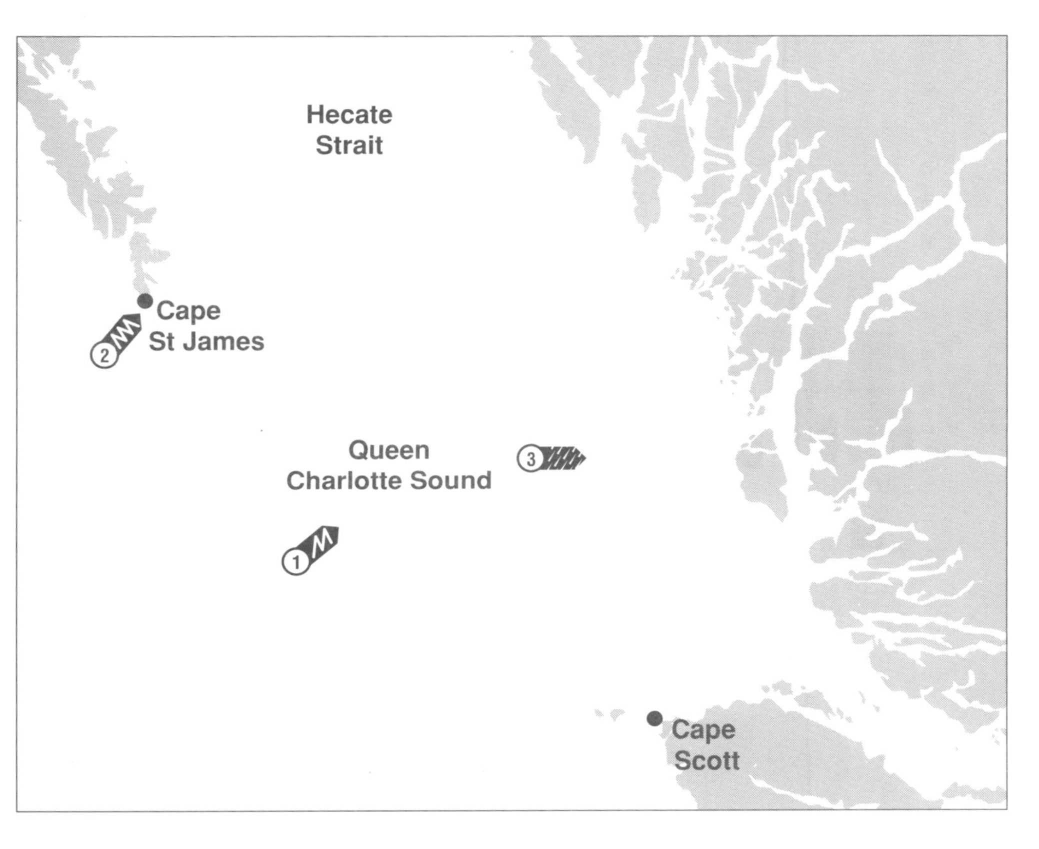

1. Strong winds and High waves The main hazard in Queen Charlotte Sound is from very strong winds and high seas associated with coastal lows. The worst conditions usually occur with the westerly winds just after the low and front have moved across the coast. Storm force winds with gusts to 90 knots and seas of 10 to 12 metres can accompany such a storm.

2. Steep waves Southerly and southwesterly wind waves and swell waves steepen on the tidal currents off the Kerouard Islands. Waves may break producing hazardous conditions especially when strong winds counter tidal streams. Cape St. James' reports of winds from the southeast through south to northwest are representative of the area.

3. Fog Sea fog is prevalent from August to October and often moves west toward the mainland coast. During September, visibilities can occasionally fall to 2 miles or less.

Winds – Cape St James

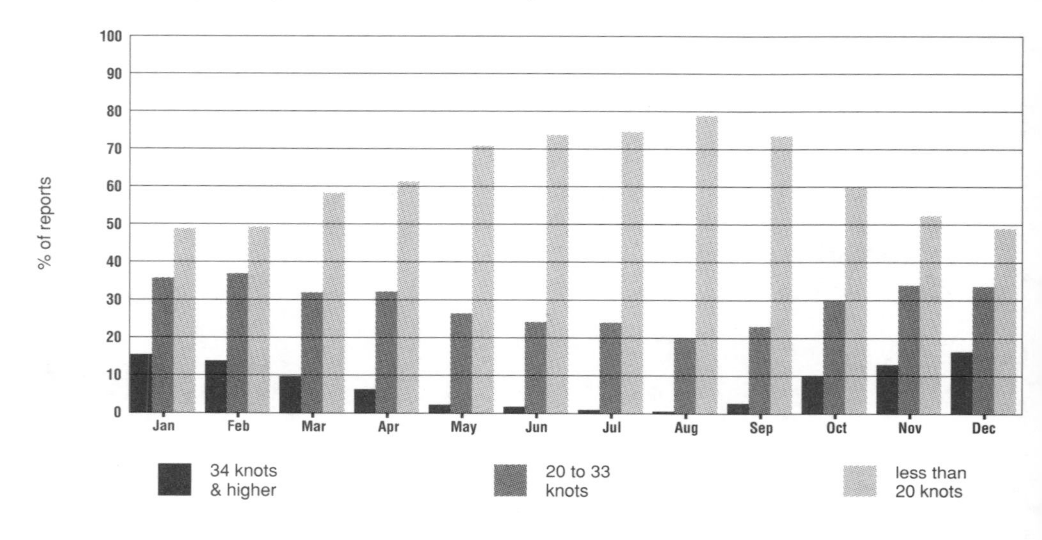

General Conditions

Winter storms producing extreme winds and high waves are the principal hazards in exposed areas of the Central Coast. Coastal lows crossing the Charlottes with an associated frontal system produce the worst conditions.

Close to the coastline, seas interact with strong tidal currents. As the waves steepen and break they may swamp or broach smaller vessels. The passage between the Scott Islands and Nahwitti Bar and the route from Cape Caution to Cape Calvert are notorious in this respect.

Cold outbreaks produce outflow winds reaching storm force in many of the mainland inlets. These winds produce extensive sea spray and spray icing. The report from Cathedral Point in Burke Channel will reflect outflow conditions.

Cold air outbreaks usually end with a Pacific frontal system moving onto the coast. As the storm approaches, southeasterly winds will spread rain into the western entrances of the inlets but snow will remain at the heads of inlets. In between, freezing rain or mixed rain and snow may be expected. Radar effectiveness is greatly reduced in the mixed rain and snow.

Cape St. James should be monitored as an early indicator of wind changes for the Central Coast. When the winds at Cape St. James increase to gale force, (for example) expect similar winds to arrive on the central coast up to 24 hours later. Likewise, when Cape St. James shows continued southwest winds after the passage of a winter frontal system, be forwarned of high southwest winds and seas spreading to the Central Coast. The southwest seas are often quite high and can be very dangerous when they funnnel into the channels of the mainland coast. These conditions will be reported by the West Sea Otter Group buoy shortly before their arrival on the coast.

In summer, sea breezes generally occur in most of the mainland inlets and fjords. Up-inlet winds begin during the morning and increase to 20 to 25 knots in the afternoon. Winds decrease in the early evening. In some areas weaker but gusty outflow drainage winds develop at night.

Visibility in the Central Coast area is best in the spring. In winter, visibility is reduced by precipitation and fog accompanying winter storms. In late summer and fall, sea fog spreads over the area from the west and occasionally lowers visibilities to less than 2 miles.

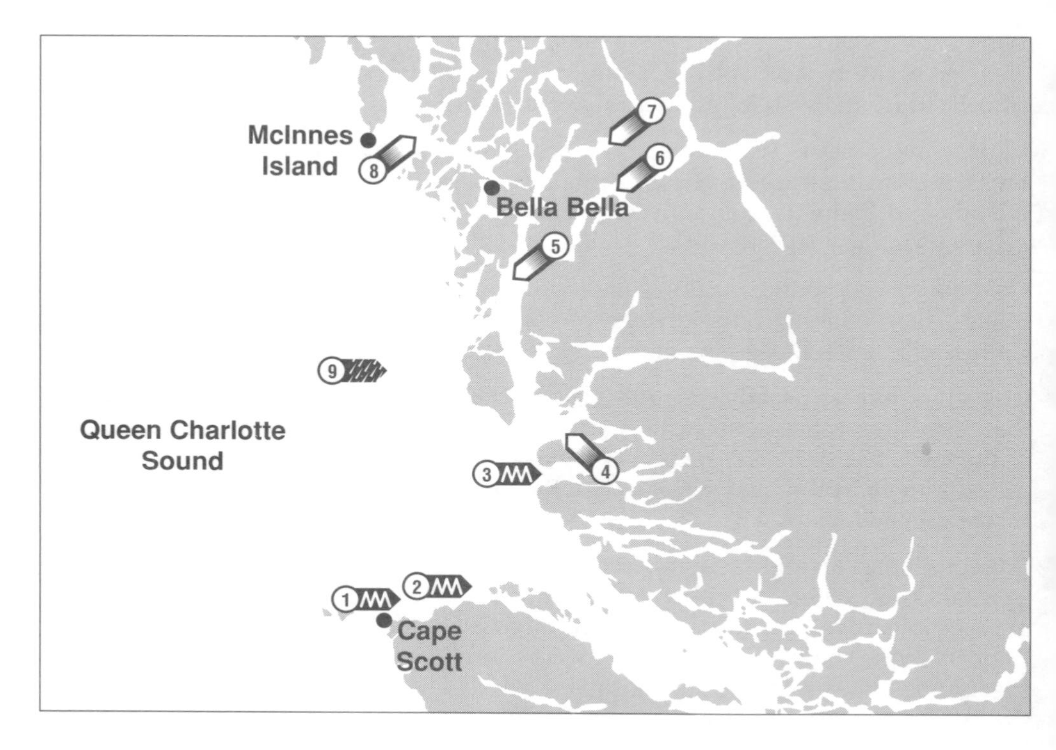

1. Steep waves Strong tidal currents from 1.5 to 3 knots are present around the Scott Islands and east to Cape Sutil, producing dangerous rips and overfalls. Westerly swell and wind waves steepen and break on these currents compounding the danger. Vessels should avoid this area in heavy weather.

2. Steep waves Nahwitti Bar in the entrance to Goletas Channel is one of the most difficult passages in the coastal waters. Westerly waves and swell from Queen Charlotte Sound steepen and break on the 5 knot ebb tides over the bar producing very confused seas.

3. Steep waves Westerly winds and waves in Smith Sound produce steep, breaking seas. Conditions between Cape Caution and Calvert Island are particularly dangerous in storms.

4. Strong winds Gusty winds are prevalent in Goose Bay and Rivers Inlet when strong southeasterlies develop. Air cascades down from the mountains at the head of the bay producing gusty conditions.

5. Strong winds Outflow winds during autumn and winter blow down the mountain valleys into harbours and over the Inner Passage. When a storm approaches after a cold arctic outbreak, south winds in Fitz Hugh Sound will meet northeast outflow winds from Burke and Dean Channels to create chaotic wind and sea conditions. Freezing rain is likely to occur further up Fisher Channel. Freezing sea spray may occur if air temperatures are below –5° Celsius.

6. Strong winds Outflow winds exceeding 60 knots are known to occur in Burke Channel during arctic outbreaks. Short, steep waves and blowing spray occur in these winds which can persist for several days. When air temperatures are cold, less than –5° Celsius, freezing spray can occur (see Icing diagram on page 75). The Cathedral Point report is representative of outflow winds along Burke Channel.

7. Strong winds Outflow winds in Dean Channel and tributaries can be severe, reaching storm force during arctic outbreaks. Sea spray icing can be heavy.

8. Strong winds West to southwest winter winds are funnelled into Seaforth Channel and other east-west channels along the Central Coast, producing very strong winds in the inlets. Rough seas are generated on ebbing tidal currents. McInnes Island Lighthouse will report these southwest winds.

9. Sea fog is prevalent from August to October and can persist for several days in a row until stronger, dryer winds flush it out of the area.

During hot summer weather, strong up-inlet winds and whitecapping waves build up in mainland fjords.

Winds–Egg Island

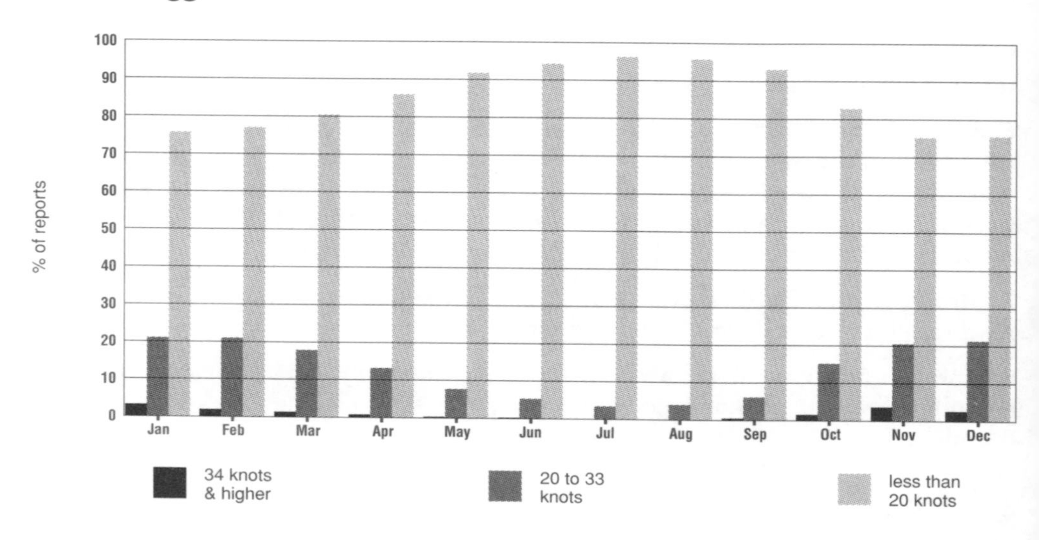

General Conditions

The worst weather hazards in Hecate Strait occur in conjunction with storms during the fall and winter months. Southeast gales are frequent, as frontal systems cross the coast. Storm force winds, which are reported less frequently, generally occur when a coastal low moves directly over the area.

Sea states vary from storm to storm. Near Bonilla Island waves can build to 6 metres whenever there are sustained gale force winds for several hours. Once or twice a winter, gale force southeasterly winds will produce wind waves which, when combined with southerly swells, build the seas to near 8 metres over northern Hecate Strait. Over the southern part of the Strait, where there is greater exposure to the swells from the Pacific, seas of 8 to 10 metres are not that uncommon during the winter.

Shallow water depths over large sections of Hecate Strait allow short steep waves to develop quickly. This makes transiting across the Strait quite dangerous for smaller vessels. Because of the speed that the winds and seas can change, it has been said that Hecate Strait is the fourth most dangerous body of water in the world.

In many of the coastal areas on both sides of Hecate Strait, southeasterly gales and waves interact with currents and bathymetry to increase the dangers for small vessels. Usually, when waves run into a countering current they steepen and sometimes break. Navigating these areas in high winds with reduced visibility is particularly hazardous.

In winter, during an arctic outbreak, gale to storm force northeasterly winds blow through the mainland inlets, including Douglas Channel. The winds decrease and become more northerly as they move out into the open water of Hecate Strait. The cold and dry arctic air picks up moisture as it crosses the Strait bringing snow showers and reduced visibilities to the east coast of the Queen Charlotte Islands. The strong winds and cold air temperatures in the inlets create rough water with freezing spray in many of the principal channels. Many locations, sheltered by an island or away from the opening of an inlet may not experience these strong winds.

In summer, northwesterly winds prevail, with speeds occasionally up to 35 knots. South to southeast winds occur in advance of weaker summer frontal systems.

During the summer, the sea breeze is a dominant factor controlling winds in the inland waterways. At about 10 a.m. winds begin to blow up the channels, and often rise to 20 to 25 knots during the afternoon. Winds are generally light at night. In some areas, particularly when snow is present at higher elevations bordering the channel, drainage winds may occur at night bringing sudden gusty conditions with speeds up to 20 knots.

The poorest visibilities, resulting from fog and snow, occur from September through February. Some sea fog also reduces the visibility in the fall, primarily in the southern part of the area. The remainder of the Strait is protected from ocean fog by the Queen Charlotte Islands. The best visibilities are found in the summer months.

Lighthouse reports are a valuable source of information on overwater winds and waves. Bonilla Island is strategically situated to give representative northwest and southeast winds over the central section of the Strait. Reports from Rose Spit give a good indication of the winds along the northern shoreline, but often the winds over the water are even stronger than the light station's reports.

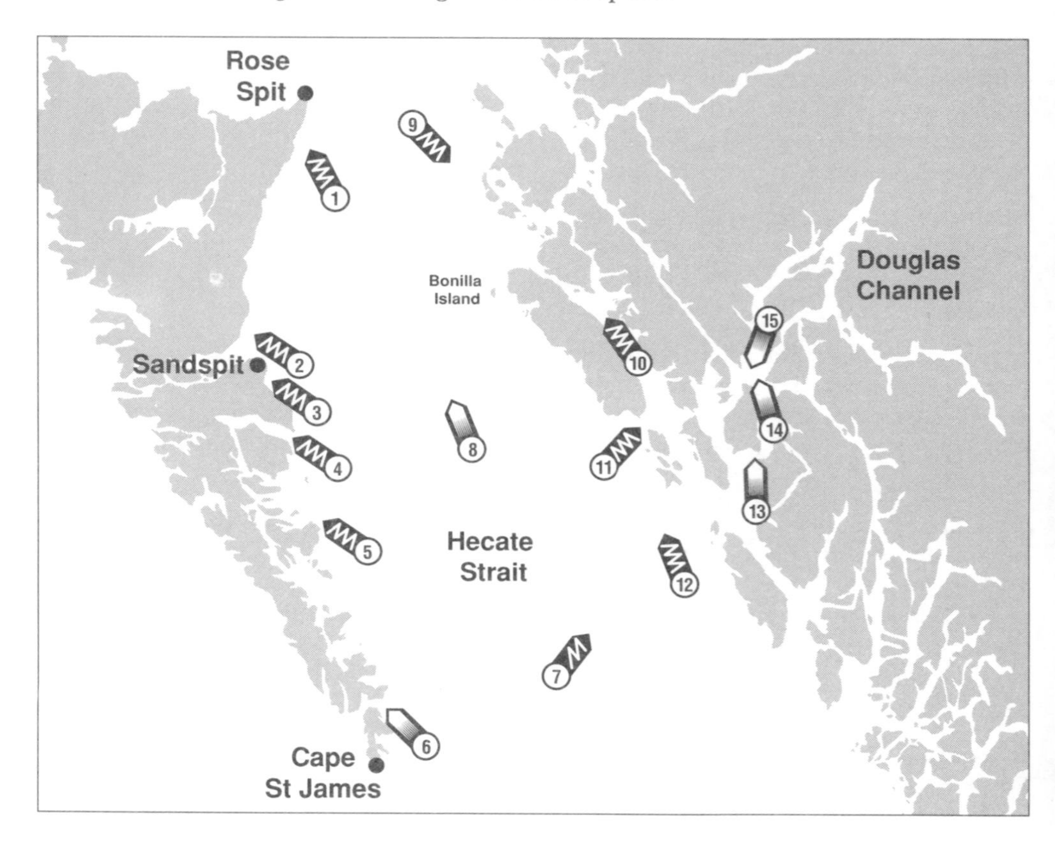

1. Steep waves Southeast gales and heavy seas in Hecate Strait affect shallow areas from Cape Ball to Rose Spit ("The Hook"). Waves steepen and break on shallow bars and on tidal currents. With eratic patterns of rocks and shifting sandbars along the east coast of Graham Island, it is often safest to remain about five miles off the coast.

2. Steep waves Southeast winds and waves produce hazardous conditions over the bar into Skidegate Inlet as waves shoal in shallow water. Waves will steepen and break over the bar.

3. Steep Waves Steep breaking waves are found about 1 to 3 miles off Cape Chroustcheff, where local mariners refer to them as "large holes in the water".

4. Steep waves Southeast winds and waves interact with tidal current at the entrance to Cumshewa Inlet producing steep, breaking waves.

5. Steep waves Southeast winds and waves steepening on ebb tide currents off Lyell Island produce dangerous seas.

6. Strong winds Southeast gales across the south end of Moresby Island produce strong winds in Rose Inlet. These flow over the neck of land at the head of Rose Inlet and descend down into Carpenter Bay producing gusty winds. These may be hazardous to small craft seeking shelter in the Bay. Wind conditions at Cape St. James are representative for the area.

7. High waves Swells from the Pacific can spread into the southern section of the Hecate Strait to build the seas in a winter storm to 8 to 10 metres. Seas of this magnitude often occur during the passage of a coastal low across the Queen Charlotte Islands.

8. Strong winds and steep waves Frequent southeast gales in winter months result in 6 to 8 metre waves which steepen on flood-tide currents. Winds and waves are usually associated with the passage of the cold front from a Gulf of Alaska low.

9. High waves Westerly winds and swell from Dixon Entrance penetrate to the coast off Porcher Island.

10. Strong winds and high waves Southeasterly winds funnelling through Principe Channel can build up quite high seas near Anger Island.

11. Steep waves Heavy seas under southerly winds produce steep, breaking waves on ebb tides through Otter Passage and Langley Passage.

12. Steep waves Southerly winds can build heavy seas along the coast from Aristazabal Island to Banks Island. Waves interacting with local tidal currents steepen to give hazardous conditions.

13. Strong winds Southerly winds are funnelled by local mountains, and descend into Whale Channel at Bernard Harbour. Gusty conditions result.

14. Strong winds When a storm begins to approach the B.C. coast, strong northeast winds blow down the entire length of Douglas Channel. As the storm advances, southeast winds develop over Hecate Strait and along Whale and Grenville Channels. These inflow and outflow winds converge at Wright Sound producing chaotic conditions.

15. Strong winds Arctic outbreaks produce gale-to-storm-force winds in Douglas Channel. Sea spray icing and waves pose a hazard to vessels.

Winds – Bonilla Island

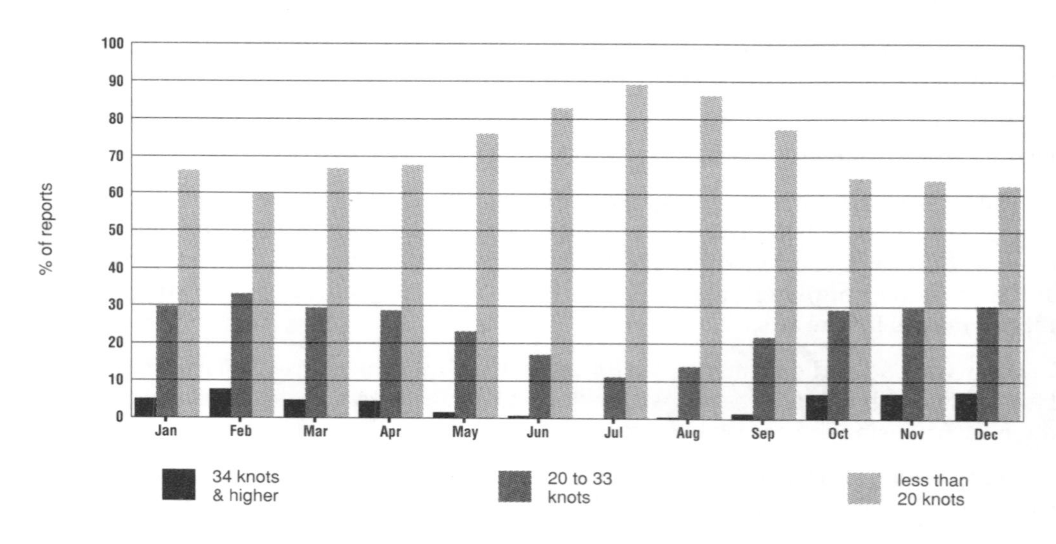

General Conditions

Dixon Entrance East marine area is well protected to the southwest by Graham Island and on the north by the rugged islands of the Alaska Panhandle. Severe weather is produced by southeasterly gales that build ahead of frontal systems from storms moving onto the coast. Winds are channelled up Hecate Strait giving heavy seas between Rose Spit and Triple Island. Strong westerly winds and waves often build behind storms with the formation of a high pressure ridge over the Gulf of Alaska.

The winds near the mainland coast are often reluctant to shift into the south or southwest after a frontal passage. They tend to remain from the southeast until the ridge builds strongly enough behind the front for northwest winds to spread into the area.

In winter months, cold arctic outbreaks are common. Northeasterly gales blow down the full length of Portland Inlet and other mainland fjords affecting the marine area. As the cold air continues over the water it picks up enough moisture to produce snow showers which reduce the visibility. The strong, cold winds can also produce freezing sea spray.

In summer, as the Pacific high pressure area extends northward, the most frequent winds in Dixon Entrance East are from the west.

Fog is prevalent in summer, occasionally reducing visibilities below 1 mile.

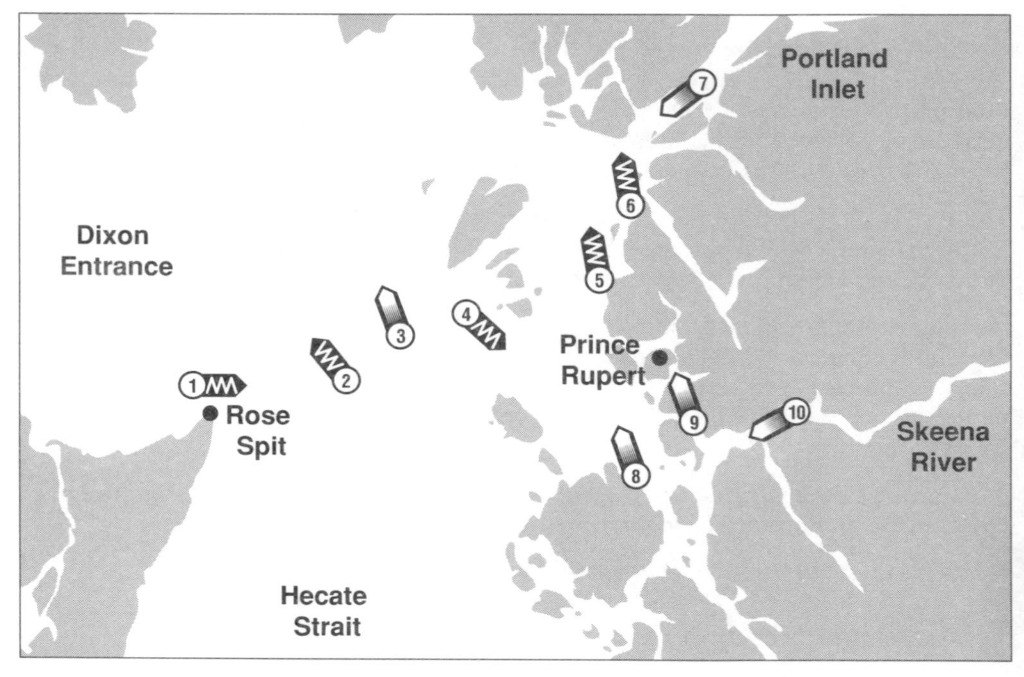

1 Steep waves Tidal currents are particularly strong north of Rose Spit, up to 3.5 knots. Near the transition from shallow to deep water strong tidal rips and heavy overfalls occur. These are hazardous at all times but become very dangerous under strong winds or in heavy seas. Westerly swell will interact with the currents producing steep, breaking waves.

2. Steep waves Southeast gales in Hecate Strait can produce seas of 5 to 7 metres, occasionally reaching 8 metres in height. These waves steepen as they shoal and break on the shallow flats east of Rose Spit. The steep waves in shallow water can capsize small vessels.

3. Strong winds Southeast gales are frequent in the winter months, October to April. Wind reports from Rose Spit, Triple Island and Green Island should be used to gauge conditions.

4. Steep waves Brown Passage is exposed to westerly waves from Dixon Entrance. Waves will steepen and break on opposing ebb tide currents, ranging from 2 to 2.5 knots at peak speeds. Forecasts of prolonged gale or storm force winds and seas from the west should provide warning of hazardous conditions in Brown Passage. The forecasts can be supplemented by reports from Triple Island on wind and wave conditions.

5. Steep waves Southeasterly winds build up steep waves near Finlayson Island.

6. Steep waves Outflow winds from Portland Inlet combine with southeasterlies up Chatham Sound to produce crossing waves in Main Passage. Green Island Lighthouse and Grey Islet reports should be used for warning of dangerous conditions.

7. Strong winds and Icing During arctic outbreaks, gale force winds can be expected down Portland Inlet and over Chatham Sound. Severe freezing sea spray often results and visibilities can be reduced in sea smoke or ice fog. Reports from Green Island and Grey Islet are representative for winds and icing.

8. Strong winds and steep waves Southeast gales over Porcher Island and through adjacent channels produce gusty winds in the southern end of Chatham Sound and Refuge Bay. When southeast seas counter the currents just south of Smith Island steep waves are formed. Reports from Holland Rock should provide some guidance.

9. Strong winds and high waves Gusty winds are found in Prince Rupert Harbour during southeast gales produced by airflow over the local mountains. A heavy buildup of seas occurs in the entrance to Prince Rupert harbour with strong northwesterly winds.

10. Strong winds and icing Gale force, easterly winds blow down the Skeena estuary during periods of arctic outbreaks. Both snow showers, with reduced visibility, and freezing spray icing may occur.

Winds–Triple Island

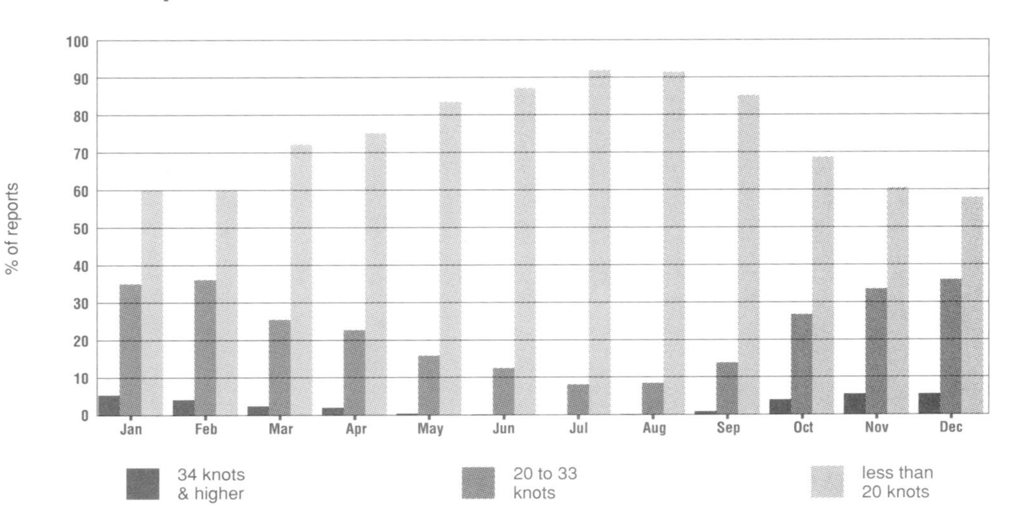

General Conditions

Dixon Entrance West is exposed to the Pacific Ocean, although the eastern half of the marine area is relatively well protected on the south by Graham Island and on the north by the rugged islands of the Alaska Panhandle. In this protected area, strong winds and waves are confined to westerly flow conditions, which often build behind storms with the formation of a high pressure ridge over the Gulf of Alaska.

Over the eastern limit of the area, southeasterly gales that build ahead of frontal systems moving onto the coast are channelled up Hecate Strait giving heavy seas off Rose Spit. Winds in the area north of Graham Island are usually backed more into the east and are often lighter than at Rose Spit.

West to northwest gales that develop behind a passing frontal system can generate combined seas of 8 to 12 metres. When the storm centre is in the northeast Gulf of Alaska, the very long fetch across the Pacific can produce swells of 5 to 6 metres with periods of 16 seconds or more.

In winter months, cold arctic outbreaks are common with northeasterly gales blowing down the length of the mainland inlets and across Dixon Entrance. As the cold air continues over the water it picks up moisture and produces snow showers, which reduce visibility. Freezing spray occurs with the strong outflow winds if the air temperatures are cold enough. (See diagram p 75)

In summer, as the Pacific high pressure area extends northward, the most frequent winds in Dixon Entrance are from the west. But even when the winds are light there is almost always some long swell rolling in from the Pacific. Seas rarely fall below 1 to 2 metres even in summer. The shallow water over Learmouth Bank enhances whatever seas are present.

Fog is prevalent in summer, occasionally reducing visibilities below 1 mile.

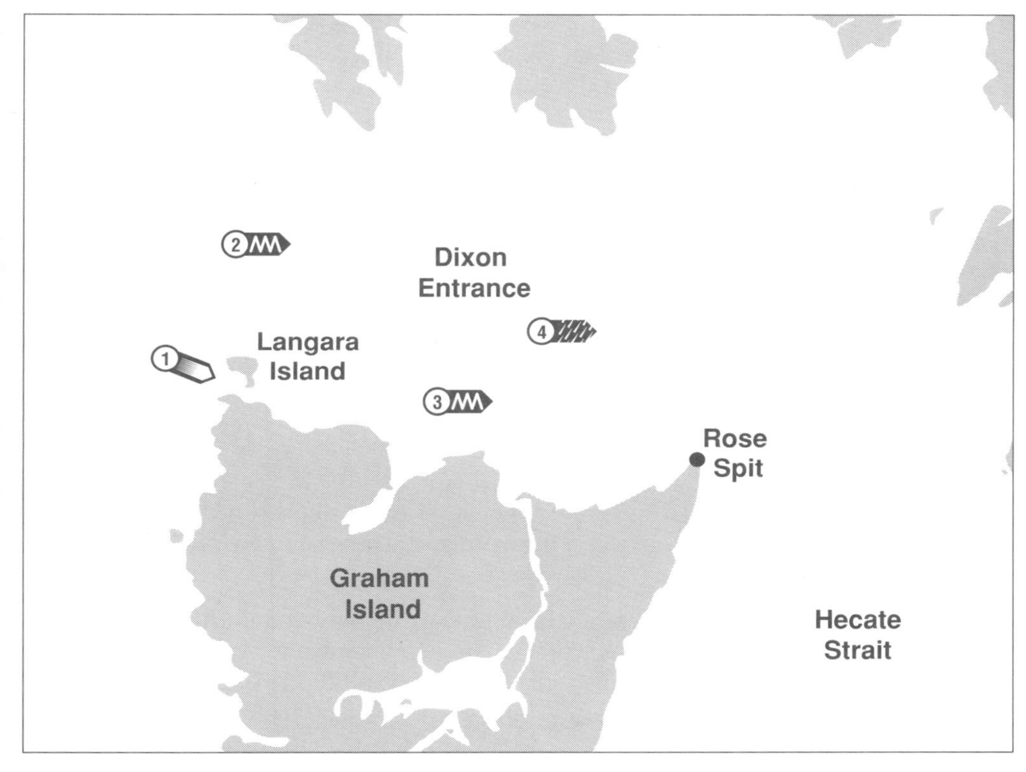

1. Strong winds and steep waves Funnelling of westerly winds in Parry Passage increases wind speed in the channel. Westerly winds and waves countering the ebb tide create dangerous conditions on an eastward passage. Wind and wave reports from Langara and the Dixon Entrance West bouy should be used for guidance.

2. Steep waves Westerly seas flowing over Learmouth Bank create dangerous overfalls and rips. An ebb tidal current opposing the westerly seas will make conditions even worse for small craft.

3. Steep waves Tidal currents set along the north shore of Graham Island range up to 2.5 knots near Langara Island. Westerly waves will steepen on ebb currents giving hazardous conditions for small vessels.

4. Fog Sea fog is prevalent over Dixon Entrance in summer months.

General Conditions

The major weather hazard along the west coast of the Queen Charlotte Islands is strong winds and high seas during winter storms. Southeast winds are most common in winter but strong winds are found from most directions.

During the summer, the prevailing winds shift to northwest under the influence of the Pacific Ocean high pressure system. Strongest winds are most frequently out of the northwest.

Gusty winds are hazardous in some of the longer inlets and sounds on Moresby Island. In a southeasterly flow, winds descending from the mountains will produce gusty conditions in the inlets.

During fall and winter, combined seas range from 8 to 12 metres in storms. Swells of 3 to 5 metres in height, with periods ranging from 16 to 22 seconds, are prevalent along the coast in the wake of storms that have moved inland. Swell heights diminish slightly during spring and are at their lowest in summer (1 to 2 metres).

In winter, during deep arctic outbreaks, cold air can envelop the Queen Charlotte Islands. Easterly outflow winds blow down west coast inlets but do not generally attain the same strength as found along the mainland. These squally drainage winds are called **williwaws**.

Lowered visibility in fog occurs chiefly from June to September and from December to February. Frequently the visibility is worst in August when sea fog can reduce visibility to 2 miles or less. Winter fog results from heavy precipitation accompanying winter frontal storms. In the summer, sea fog forms when warm air from lower latitudes moves over colder coastal waters.

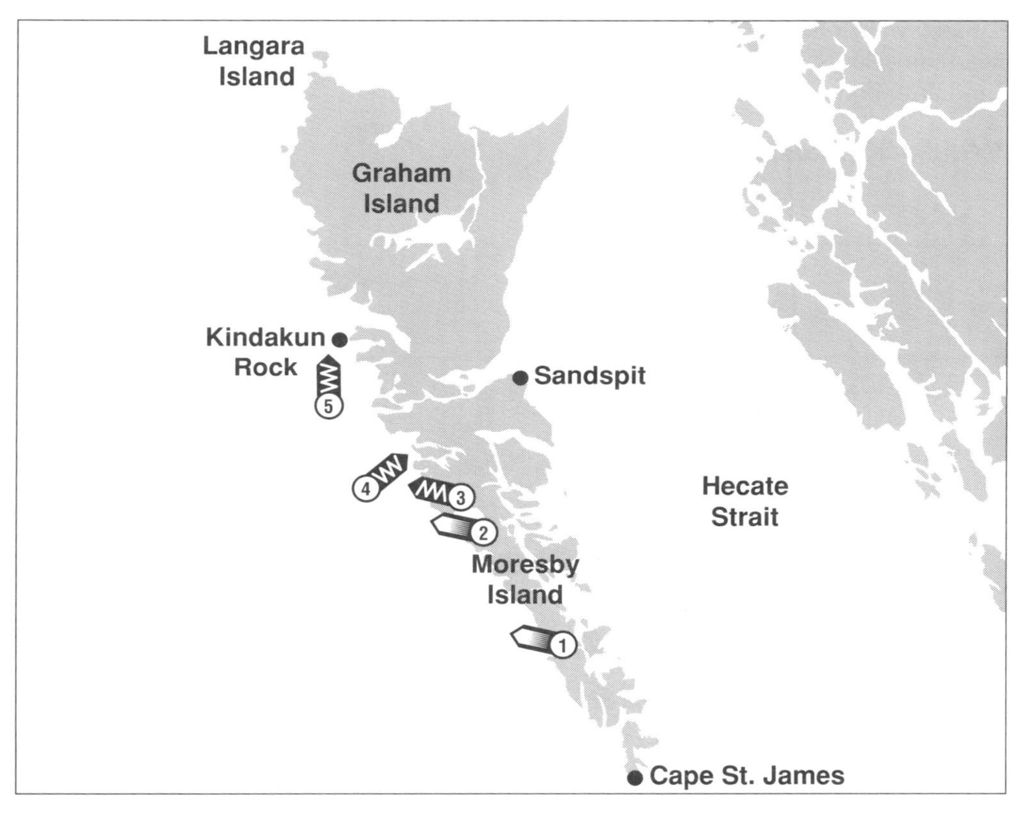

1. Gusty winds Southeast gales in Hecate Strait blow across Moresby Island and down into Gowgaia Bay producing gusty winds. Cape St. James wind reports are representative of southeast flows likely to produce unsettled conditions in the Bay.

2. Strong winds Southeast gales in Hecate Strait cross Moresby Island and blow down from the surrounding mountains to give gusty winds in Tasu Sound. Forecasts of southeast winds ahead of fronts approaching the coast combined with the Cape St. James Lighthouse reports will provide some warning of conditions expected in Tasu Sound.

3. Steep waves Westerly winds and waves become hazardous as waves steepen on tidal currents ebbing out of Kootenay Inlet. This is a general situation in many inlets along the coast.

4. Steep waves West to southwest waves surge into many inlets on the West Coast of the Charlottes causing confused steep waves.

5. Steep waves South to southwest winds and sea encounter opposing currents around headlands south of Rennell Sound. Steep, breaking waves form on tidal currents off Kindakun Point and Hunter Point.

Winds – Kindakun Rock

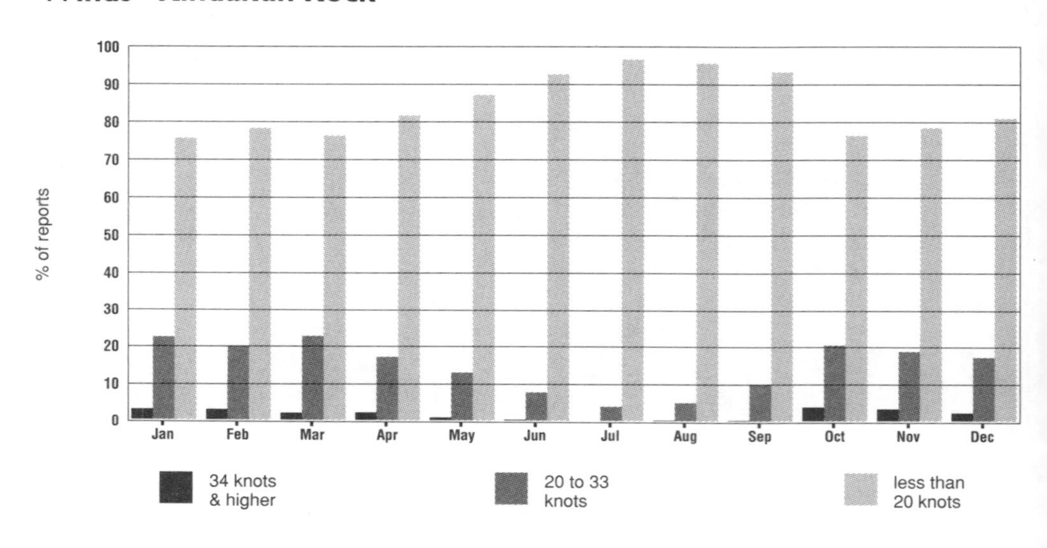

APPENDICES

Glossary of Commonly Used Terms

Backing — A counterclockwise change in wind direction (for example, from north to northwest to west, from east to northeast to north, etc.). Opposite to veering.

Bathymetry — The shape of the sea bed.

Beam Seas — Seas moving toward the side of the vessel.

Breaking Wave — A wave which has become unstable because of steepness or because the crest has overtaken the trough in shallow water.

Cold Front — A boundary zone separating cold and warm air where the cold air is displacing the warmer air.

Complex Low — A region of low pressure in which there are several main low pressure centres. This term is often used when one low which was the dominant low is weakening but another centre is developing.

Deep Low — A rather subjective term used to describe the central pressure of a low centre (usually when it is about 975 millibars or less). Often has winds of gale to storm force around the low.

Developing Low — A low in which the central pressure is decreasing with time. Winds would normally increase as the low deepens.

Dew point — The temperature at which air becomes saturated, allowing condensation of water vapour to occur.

Duration — The length of time the wind persists without changing direction or speed.

Fetch — The distance which wind blows across water from a constant direction and with constant speed.

Filling Low — A low centre in which the central pressure is increasing with time, i.e. the low is gradually weakening.

Following Seas — Seas moving in the same direction as the vessel.

Front — The line of separation between cold and warm air masses. Usually a front is associated with a change in the weather such as a wind shift, pressure rise, clearing weather, temperature change. Types of fronts: (1) warm front—

Glossary of Commonly Used Terms

APPENDIX I

	warmer air displacing colder air and (2) cold front—colder air displacing warmer air.
Gale	A sustained wind speed of 34 to 47 knots.
Gale Warning	A warning to mariners of impending winds from 34 to 47 knots.
Ground Swell	Waves that have travelled from a distant weather system where they were created. They are characteristically long crested (ie. have a long period), low and regular waves.
Gust	A sudden, brief increase in wind speed, generally lasting less than 20 seconds.
High	A region of high pressure. Winds blow clockwise around the centre. A high is usually associated with good weather.
Hurricane Force Wind	Winds which are 64 knots or more.
Hurricane Force Wind Warning	A warning to mariners of impending winds of 64 knots or more.
Isobar	Line on a weather map joining points of equal pressure.
Knot	A unit of wind speed equal to one nautical mile per hour.
Land Breeze	A small-scale wind circulation set off when the air temperature over the sea is warmer than that over the adjacent land. The land breeze develops at night and blows from the land out to sea. Its counterpart is the sea breeze. (See Land and Sea breezes, p. 52.)
Light Winds	Wind speeds ranging from 1 to 11 knots.
Locally Stronger Winds	Winds in many small areas are expected to be stronger than the general wind forecast for the marine area. Areas where this is likely to occur can be identified in the section on local marine hazards.
Low	A region of low pressure. Winds flow counterclockwise around the low centre. A low pressure centre is usually a storm centre accompanied by precipitation and strong winds.

Glossary of Commonly Used Terms

Millibar — A unit used to measure barometric pressure. A conversion table between inches of mercury, millibars and kilopascals is shown in Appendix 3.

Moderate Winds — Winds with speeds in the range 12 to 19 knots.

Outflow Winds — Winds that blow down fjords and inlets from the land to the sea. When cold arctic air flows from the interior of the province onto the coast, the wind speeds in mainland inlets can reach over 60 knots.

Overfalls — Areas of turbulent water caused by strong currents moving over submerged ridges or shoals.

Period of waves — The time, in seconds, it takes for successive wave crests (or troughs) to pass a fixed point.

Pressure Gradient — Difference in pressure between two points divided by the distance between them. The greater the difference in pressure between the same two points, the greater the wind. The closer the isobars are together on a weather map, the greater the pressure gradient.

Quartering Sea — Seas moving onto the ship's quarter at about an angle of 45° to its heading.

Refraction of waves — The change in direction of movement and size of sea waves which encounter shallow water. Waves tend to move (refract) toward the shallower water.

Ridge — An elongated area of high pressure.

Sea Breeze — A small-scale wind circulation set off when the air temperature over the land is greater than that over the adjacent sea. The sea breeze develops during the day and blows from the sea to the land. Its counterpart is the land breeze. (See Land and Sea breezes, p. 52.)

Seas — Combined wind waves and swell. Mariners often use the term to mean wind waves only, which is not the same meaning used in this manual. Forecasters will usually use the term "combined sea" instead to avoid this ambiguity.

Sea State Forecast — Forecast of height of the combined wind wave and swell given in metres.

Glossary of Commonly Used Terms

Shallow Water — Water depths less than or equal to one half of the wavelength of a wave. Therefore, water may be "shallow" for some waves, but not for others.

Shoalling — The process whereby waves coming into shallow water become shorter and steeper.

Significant Wave Height — Average height of the highest one-third of the waves present.

Squall — An increase in the wind of longer duration than a gust. Usually it is associated with showers from a cumulonimbus cloud. A squall generally lasts several minutes.

Stable air — A non-turbulent state of the atmosphere which occurs when warm air is over cold air. The air does not easily move up or down but is forced to remain on the horizontal plane.

Steep Waves — A wave which has a steep slope or angle to the waves. It is almost the opposite to a long rolling wave.

Storm Force Winds — Sustained winds which are 48-63 knots.

Storm Warning — The wind warning that is issued to mariners when winds are expected to be 48-63 knots.

Strong Winds — Winds speeds ranging from 20 to 33 knots.

Swell — Long waves formed from a distant storm, more regular in appearance than wind waves and no longer growing in height.

Tidal Rip — A heavy boil on the sea surface often accompanied by breaking waves. Rips are produced by strong tidal currents moving against a wind generated wave, or by a rapid flow of water over an irregular sea bottom.

Tidal Wave — The daily movement of water produced by the forces of the sun and moon. Tides move as a long, shallow water wave with a period often between 12 to 24 hours. A tsunami and storm surge are often wrongly called a tidal wave.

Topography — The shape of the land.

Trough — An elongated area of low pressure, often associated with a wind shift and showery weather.

Glossary of Commonly Used Terms

Unstable air — A turbulent state in the atmosphere, often caused by cold air moving over warm air. Unstable air has a tendency to move more up or down in order to balance the unnatural, or unstable, air temperature profile.

Veering — A clockwise change in wind direction, such as from south to southwest.

Warm Front — A boundary separating cold and warm air masses. The warmer air is overtaking the colder.

Waterspout — A small whirling storm over water which is spawned from the base of a thunderstorm. It is similar to but generally not as severe as a tornado.

Wave length — The length between two crests (or troughs) of the wave.

Wind Waves — Waves generated by the local wind.

Cold Water Survival

Hypothermia (or lowered deep-body temperature) can be life-threatening to any mariner exposed to B.C.'s cold, coastal waters. This fact adds to the importance of making wise weather decisions that could affect boating safety. The following information on cold water survival is reprinted from *Sailing Directions, British Columbia Coast, North Portion* (Fisheries and Oceans, 1983).

It should be noted that while the following information is correct the actual ability of an individual to survive in cold water does vary considerably from person to person. There are several known cases in which an individual has survived in cold water for extended periods of time.

Without special clothing such as an immersion suit or Personal Flotation Device (PFD) with good thermal protection, even a short period of immersion in cold ocean water causes hypothermia, which can be fatal.

In cold water, the skin and external tissues cool very rapidly but it takes 10 to 15 minutes before the temperature of the heart, brain and other internal organs begins to drop.

Intense shivering occurs in an attempt to increase the body's heat production and counteract the large heat loss.

Once cooling of the deep body begins, body temperature falls steadily and unconsciousness can occur when the deep-body temperature drops from the normal 37°C to approximately 32°C. When the body core temperature cools to below 30°C, death from cardiac arrest usually results.

In a water temperature of 5°C, persons without thermal protection become too weak to help themselves after about 30 minutes. Even if rescued, the chances of survival after an hour of immersion are slim.

The body, in almost all weather conditions, cools much faster in water than in air, thus the less body surface submerged, the better. The parts of the body with the fastest heat loss are the head and neck, the sides of the chest and the groin. To reduce body heat loss, protect these areas.

Two ways of reducing heat loss are:

(a) HELP (Heat Escape Lessening Position) arms held tight against the sides, ankles crossed, thighs close together and raised;

(b) Huddle two or more persons in a huddle with chests held close together. To use these methods successfully a person must be wearing a lifejacket or PFD.

Survival time is greatly increased by wearing clothing that gives thermal protection, including a hood to prevent heat loss through the head.

APPENDIX 2

Cold Water Survival

Do not swim to keep warm as this causes heat to be lost to the cold water due to more blood circulation to the arms, legs and skin. Tests show that a person in a lifejacket cools 35 per cent faster when swimming than when holding still.

If you have no lifejacket or other flotation, tread water remain as still as you can, moving your arms and legs just enough to keep the head out of the water.

Although tests show that the heat loss is 35 per cent faster when treading water than when holding still in a lifejacket, this is much better than the "drownproofing" technique with which the heat loss is 82 per cent faster, mainly due to the head (a high heat loss area) being periodically submerged.

Rewarming after mild hypothermia

If the casualty is conscious, talking clearly and sensibly and shivering vigorously, then:

(a) get the casualty out of the water to a dry sheltered area;
(b) remove wet clothing and if possible put on layers of dry clothing; cover the head and neck;
(c) apply hot, wet towels and water bottles to the groin, head, neck and sides of the chest;
(d) use electric blankets, heating pads, hot baths or showers;
(e) use hot drinks but never alcohol as it does not warm a person.

Rewarming after severe hypothermia

If the casualty is getting stiff and is either unconscious or showing signs of clouded consciousness such as slurred speech, or any other apparent signs of deterioration, immediately (if possible) transport the casualty to medical assistance where aggressive rewarming can be initiated.

Once shivering has stopped, there is no use wrapping casualties in blankets if there is no source of heat as this merely keeps them cold; a way of warming them must be found quickly. Some methods are

(a) put the casualty in a sleeping bag or blankets with one or two warm persons, with upper clothing removed;
(b) use hot, wet towels and water bottles as described previously;
(c) warm the casualty's lungs by mouth-to-mouth breathing.

Caution Warm the chest, groin, head and neck but not the extremities of the body; warming the extremities can draw heat from the area of the heart, sometimes with fatal results. For this reason do not rub the surface of the body. Handle the casualty gently to avoid damaging the heart.

Metric Conversion Tables

Table for converting metres to feet for sea state forecasts.

1 metre	3 feet	**8 metres**	26 feet
2 metres	7 feet	**9 metres**	30 feet
3 metres	10 feet	**10 metres**	33 feet
4 metres	13 feet	**12 metres**	40 feet
5 metres	16 feet	**15metres**	49 feet
6 metres	20 feet	**20 metres**	66 feet
7 metres	23 feet	**25 metres**	82 feet

Table showing various pressure conversions

inches	millibars	kilopascals	inches	millibars	kilopascals
28.05	950	95.0	**29.85**	1010	101.0
28.20	955	95.5	**29.95**	1015	101.5
28.35	960	96.0	**30.10**	1020	102.0
28.50	965	96.5	**30.25**	1025	102.5
28.65	970	97.0	**30.40**	1030	103.0
28.80	975	97.5	**30.55**	1035	103.5
28.95	980	98.0	**30.70**	1040	104.0
29.10	985	98.5	**30.85**	1045	104.5
29.25	990	99.0	**31.00**	1050	105.0
29.40	995	99.5	**31.15**	1055	105.5
29.55	1000	100.0	**31.30**	1060	106.0
29.70	1005	100.5	**31.45**	1065	106.5

Table showing various wind speed conversions

knots	mph	km/h	knots	mph	km/h
10	11	18	**55**	63	102
15	17	28	**60**	69	111
20	23	37	**65**	75	120
25	29	46	**70**	81	130
30	35	56	**75**	86	139
35	40	65	**80**	92	148
40	46	74	**85**	98	157
45	52	83	**90**	104	167
50	58	93	**95**	109	176

APPENDIX 4

Photo Credits and References

Photo credits:

Canadian Coast Guard, Search and Rescue Prevention Group, p. 55
Environment Canada, pp. 27, 41, 60, 61, 73, 77, 111
Owen S. Lange, pp. 17, 35, 67, 69 (lower), 71
Paul LeBlond, p. 65
Bob Loveless, pp. 69 (upper), 70 (lower)
Duncan McDougall F.R.P.S., pp. 5, 70 (upper)
Seaconsult, Vancouver, p. 74
Ministry of Crown Lands, Surveys and Resource Mapping Branch, Victoria, B.C., p. 94

Reference List:

Canada Ministry of Supply and Services, Canadian Coast Guard. ***Small Fishing Vessel Safety Manual.*** Ottawa. 1989.

Canada. Scientific Information and Publications Branch. Fisheries and Oceans. ***Sailing Directions, British Columbia Coast.*** Ottawa, 1984.

Kinsman, B. ***Wind Waves: Their Generation and Propagation on the Ocean Surface.*** Englewood Cliffs, N.J.: Prentice-Hall, Inc., 1984.

Lange, Owen, ***The Wind Came All Ways.*** Environment Canada, Vancouver, B.C., 1998. ISBN # 0-660-17517-7 Cat No. En56-74/1998E

Lilly, Kenneth E., Jr. ***Marine Weather of Western Washington.*** Starpath School of Navigation, Seattle, WA., 1983.

Thomson, Richard E., ***Oceanography of the British Columbia Coast.*** Department of Fisheries and Oceans. Ottawa, 1981.

www.weatheroffice.com

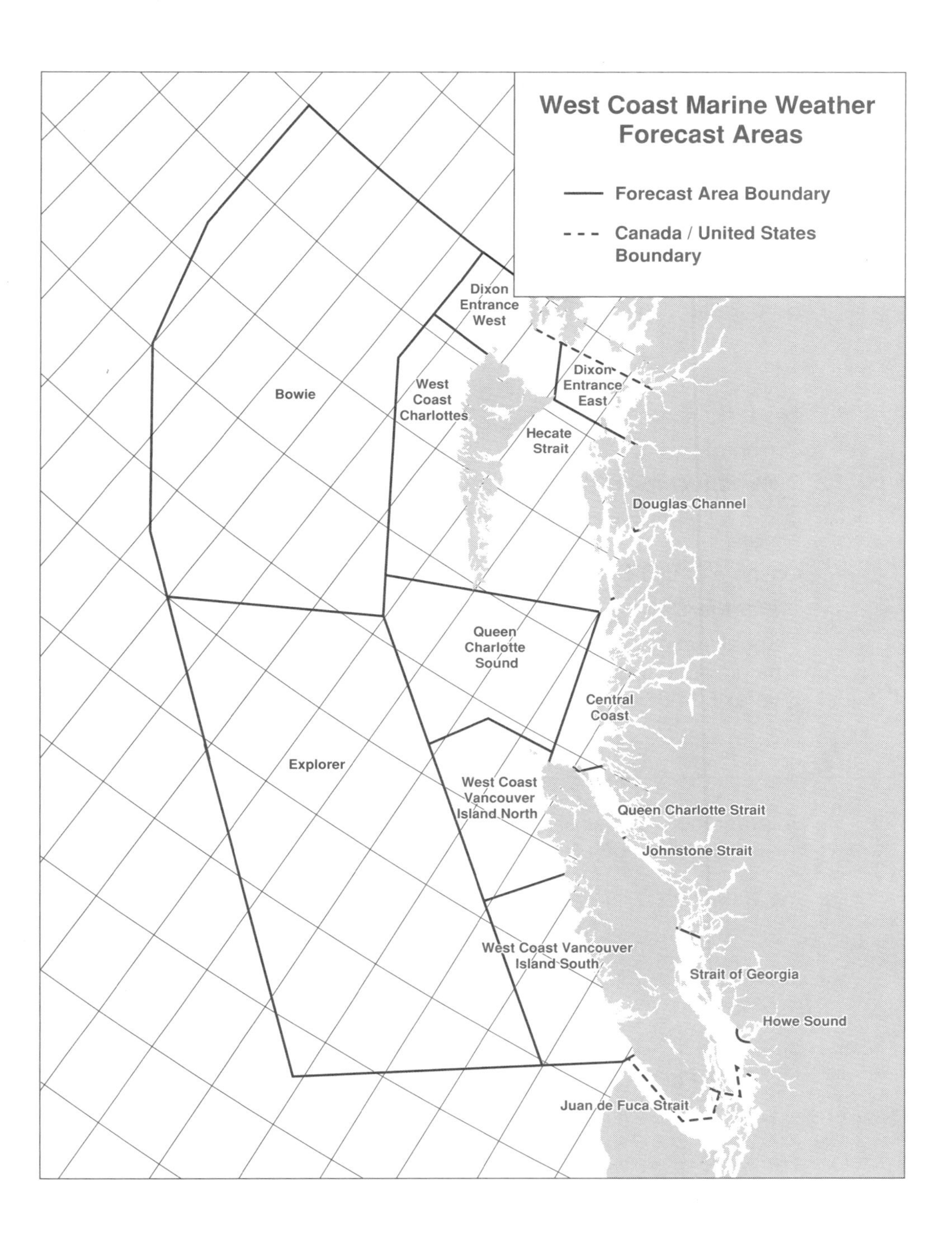
West Coast Marine Weather Forecast Areas
Forecast Area Boundary
Canada / United States Boundary
Dixon Entrance West
Dixon Entrance East
West Coast Charlottes
Hecate Strait
Bowie
Douglas Channel
Queen Charlotte Sound
Central Coast
Explorer
West Coast Vancouver Island North
Queen Charlotte Strait
Johnstone Strait
West Coast Vancouver Island South
Strait of Georgia
Howe Sound
Juan de Fuca Strait

OBTAINING MARINE SERVICES

FREE recordings

Vancouver	604-664-9010	Campbell River	250-286-3575
Victoria	250-656-2714/15	Port Hardy	250-949-7148
Nanaimo	250-245-8899	Prince Rupert	250-624-9009
Comox	250-339-5044		

WeatherPhone 604-664-9010

FREE recorded marine forecasts for Howe Sound, Strait of Georgia, Haro Strait & Juan de Fuca Strait

---

Services for a fee:

Onshore:

WeatherOne-on-One 1-900-565-6565

Talk directly to a Marine Forecaster

Onshore or OFFSHORE:

WeatherCall

Speak directly to a marine forecaster. An access phone number will be provided upon registration. To register phone 604-664-9033 or fax 604-664-9081.

To order more copies of this manual or "The Wind Came All Ways",
Call: 604-664-9360, **Fax:** 604-664-9181, or **write to:**

Commercial Services
Environment Canada
120-1200 West 73rd Avenue
Vancouver, British Columbia
V6P 6H9

For other commercial marine weather services
Phone: 604-664-9033
Fax: 604-664-9081